INTER-RATER RELIABILITY USING SAS

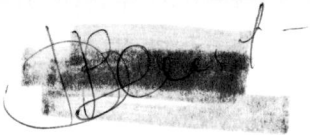

Also by Kilem L. Gwet

▶ HANDBOOK OF INTER-RATER RELIABILITY (Second Edition): *The Definitive Guide to Measuring the Extent of Agreement Among Multiple Raters* (ISBN: 978-0970806246 / 978-0970806222)

▶ HOW TO COMPUTE INTRACLASS CORRELATION USING MS EXCEL: A Practical Guide to Inter-Rater Reliability Assessment for Quantitative Data. *eBook downloadable at*:
www.agreestat.com

INTER-RATER RELIABILITY USING SAS

A Practical Guide for Nominal, Ordinal, and Interval Data

Kilem Li Gwet, Ph.D.

Advanced Analytics, LLC
P.O. Box 2696
Gaithersburg, MD 20886-2696
USA

Copyright © 2010 by Kilem Li Gwet, Ph.D. All rights reserved.

Published by Advanced Analytics, LLC . Printed and bound in the United States of America.

No part of this book may be reproduced or transmitted in any form or by any means, electronic or mechanical, including photocopying, recording, or by an information storage and retrieval system – except by a reviewer who may quote brief passages in a review to be printed in a magazine or a newspaper – without permission in writing from the publisher. For information, please contact **Advanced Analytics, LLC** at the following address :

> Advanced Analytics, LLC
> PO BOX 2696,
> Gaithersburg, MD 20886-2696
> e-mail : info@advancedanalyticsllc.com

This publication is designed to provide accurate and authoritative information in regard of the subject matter covered. However, it is sold with the understanding that the publisher assumes no responsibility for errors, inaccuracies or omissions. The publisher is not engaged in rendering any professional services. A competent professional person should be sought for expert assistance.

Publisher's Cataloguing in Publication Data :

Gwet, Kilem Li
Inter-Reliability with SAS
   A Practical Guide for Nominal, Ordinal, and Interval Data/ By Kilem Li Gwet
      p. cm.
      Includes bibliographical references and index.
   1. Biostatistics
   2. Statistical Methods
   3. Statistics - Study - Learning. I. Title.
   ISBN 978-0-9708062-6-0

Contents

Preface ix

CHAPTER
1. The SAS Solution and Its Problems 1
   1.1 Overview 1
   1.2 Number of Raters Limited to 2 2
   1.3 Agreement Coefficient Options Limited to Kappa ... 3
   1.4 The Diagonal Problem 4
   1.5 The Unbalanced-Table Problem 5
   1.6 The Ordinal Data Problem 7

2. Agreement Coefficient for 2 Raters: A Review 9
   2.1 Overview 9
   2.2 Agreement for 2 Raters & 2 Categories 10
   ▶ Kappa Coefficient 11
   ▶ Scott's π-Coefficient 13
   ▶ Bennet's S-Coefficient 14
   ▶ Gwet's AC_1-Coefficient 14
   2.3 Agreement for 2 Raters & 3 Categories or More ... 16
   2.4 Weighting Agreement Coefficients 19
   ▶ The Linear Weights 20
   ▶ The Quadratic Weights 21
   ▶ Calculating Weighted Coefficients 22

3. Kappa and the FREQ Procedure of SAS 29
   3.1 Overview 29
   3.2 Organizing Your Data 30
   ▶ The Contingency Table 32
   ▶ The Marginal Homogeneity Test 33
   ▶ The Kappa Statistics 33

3.3 Potential Data Problems........................... 37
▶ The Unbalanced-Table Problem................. 37
▶ The Diagonal Problem......................... 38

4. Weighted Kappa & the FREQ Procedure of SAS 45
4.1 Overview... 45
4.2 The Weights..................................... 46
▶ The Cicchetti-Allison Weights.................. 47
▶ The Fleiss-Cohen Weights..................... 48
▶ Meaning of the Weights........................ 50
▶ Warning.. 50
4.3 The AgreeStat_2SAS Macro 56
▶ An Example..................................... 56
▶ The AgreeStat_2SAS Parameters................ 58
▶ The Macro's Output 59
▶ Some Remarks on the Macro.................... 64
4.4 Testing Kappa for Statistical Significance.......... 66

5. Kappa for Multiple with SAS...................... 77
5.1 Introduction..................................... 77
5.2 Agreement Among 3 Raters or More: A Review.... 78
▶ Fleiss' Generalized Kappa...................... 81
▶ Conger's Generalized Kappa 83
▶ Brennan-Prediger Coefficient.................... 85
▶ Gwet's AC_1 Coefficient......................... 85
▶ Scott's Generalized Coefficient - **Kappa** · FC..... 87
5.3 The AgreeStat_3SAS Macro 88
▶ The Macro's Output 90
▶ Macro Usage: Description of Program 5.1....... 94
▶ Distribution of Raters by Subject and
   Category as Input Data....................... 99
▶ Weighting Issues............................... 104

6.	**Rater Agreement with SAS Enterprise Guide**	**105**
	6.1 Introduction	105
	6.2 Agreement Coefficients for 2 Raters	106
	▶ Testing the AgreeStat_2EG.egp Project File	106
	▶ Modifying the Input Files	111
	6.3 Agreement Coefficients for 3 Raters or More	113
	▶ Testing the AgreeStat_3EG.egp Project File	113
	▶ Modifying AgreeStat_3EG.egp's Input Files	118
7.	**Concluding Remarks**	**121**
Bibliography		**125**
Author Index		**129**
Subject Index		**131**

Preface

I wrote this book, primarily to assist researchers, and students with the calculation of various inter-rater reliability coefficients for nominal, ordinal, and interval data using the **SAS** system. The primary focus here is to show practitioners simple step-by-step approaches for organizing their rating data, creating **SAS** datasets, and using appropriate procedures, or special macro programs to obtain the final coefficients. The agreement coefficients used in this book are first briefly described before being calculated with **SAS**. I deliberately decided to avoid any formal mathematical presentation of the coefficients under investigation in this book. Instead, these coefficients are introduced using simple numeric examples to show their functionality. This approach has the advantage of presenting the methods in a clear and simple manner, allowing the reader a quick understanding of the mechanics behind the methods. Its biggest disadvantage is perhaps some loss of generality and mathematical rigor, which may not be essential for practitioners less familiar with the field of inter-rater reliability assessment. But advanced readers who want detailed discussion of the different methods may want to see Gwet (2010a).

Although, the use of **SAS** here is basic, the intent in this book is not to teach **SAS**. Nevertheless, if the user has access to **SAS** and knows how to launch it, this may be all what is needed to compute various inter-rater reliability coefficients with the techniques recommended here. However, some experience with **SAS** is definitely required to customize my (macro) programs for specific needs. To take full advantage of the techniques I recommend, it is essential to ensure that the SAS system at your disposal has a license for the IML Procedure also known as the SAS/IML software. IML stands for Interactive Matrix Language, and I used it extensively in some of the solutions I propose. You do not need to know anything about SAS/IML to use my solutions. All you need to know is whether the SAS system you are using can run IML statements.

The FREQ procedure of SAS offers the calculation of Cohen's Kappa as an option, when the number of raters is limited to 2. The introduction of this feature is without doubt a very welcome addition to the system. But I have realized that in addition to offering only Kappa as the only agreement coefficient, the use of FREQ to compute Kappa is full of pitfalls that could easily lead a careless practitioner to wrong results. For example, if one rater does not use one category that another rater has used, SAS does not compute any Kappa at all. I refer to this problem in chapter 1 as the unbalanced-table issue. Even more seriously, if both raters use the same number of different categories, SAS will produce "very wrong" results, because the FREQ procedure will be matching wrong categories to determine agreement. I refer to this issue in chapter 1 as the "Diagonal Issue." There are actually a few other potentially serious problems with weighted Kappa that I have noticed. I have clearly documented all of the problems I was able to identify, and propose a plan for resolving them.

Many analysts are introduced to the SAS system these days through Enterprise Guide. For these analysts I have devoted an entire chapter to the calculation of inter-rater reliability coefficients with SAS Enterprise Guide (EG). Examples of EG project files are presented to show how agreement coefficients would be computed for 2 raters, as well as for 3 raters or more. I am myself an old type PC SAS programmer who at first was reluctant getting into the new SAS point-and-click platform of Enterprise Guide. However, the more I got into it, the more I liked it. EG indeed makes it

Preface

easier to implement the solutions I am proposing in this book.

I used **SAS** version 9.2 when writing this book. If you are using an older version of SAS, you will still find this book useful, except perhaps chapter 6 on Enterprise Guide. There is also a possibility that in future versions of **SAS**, **SAS** Institute will correct some of the problems associated with the **FREQ** Procedure that I documented in this book. In that happens, I will revise this book to reflect those changes.

 Kilem Li Gwet, Ph.D.

CHAPTER 1

The SAS Solution and Its Problems

1.1 Overview

My book entitled "Handbook of Inter-Rater Reliability: *The Definitive Guide to Measuring the Extent of Agreement Among Multiple Raters,*" is essentially about methodology, where I discuss at length about the merits of a large number of techniques for evaluating the extent of agreement among raters. In this book, I have decided to shift my focus, from discussing the methods' merits to producing numbers. Because the focus is on production, I confined myself to presenting a non-mathematical overview of the techniques, and to concentrate on one technology, which is **SAS**. By choosing **SAS**, I do not really anticipate that a student or a researcher will purchase this software for the sole purpose of computing inter-rater reliability coefficients. **SAS** is indeed a massive and expensive system typically licensed by institutions. But many students, and professional researchers can access it through their respective institutions. **SAS** Institute, Inc., the developer of this system offers an inexpensive learning edition, which can only be of limited help for implementing the approaches I recommend in this book. The reason is that many of the solutions I recommend use the **SAS/IML** software also known as **Proc IML**, which is not included in the learning edition.

To respond to the growing demand from researchers for software products that can compute inter-rater reliability coefficients, particularly Cohen's Kappa, **SAS** now includes an option for calculating the Kappa and Weighted Kappa coefficients in the **FREQ** procedure. Actually the **FREQ** procedure does not only compute the coefficients, it also computes associated precision measures, such as the standard errors, P-values, and confidence intervals. The community of researchers who already use **SAS** can now take advantage of these features. However, the solution proposed by **SAS** with its **FREQ** Procedure carries a large number of problems. This book is an attempt to clarify and document many of them, and to propose solutions. I am now going to review what I consider as being among the most important of these problems.

1.2 Number of Raters Limited to 2

The implementation of Kappa in the **FREQ** procedure is almost entirely based on the book that Fleiss and others wrote in 2003[1], and is limited to 2 raters only. Many inter-rater reliability experiments involve 3 raters or more. Although several agreement coefficients have been proposed in the literature for multiple raters, none is yet implemented in **SAS**. However, the **SAS** Institute's support group has developed the **magree.sas** macro program that could be downloaded at:

http://support.sas.com/kb/25/addl/fusion25006_1_magree.sas.txt

This macro program is also based entirely on Fleiss et al. (2003) recommendations, many of which could be highly questionable, particularly those related to the standard errors associated with the coefficients. I have developed another **SAS** macro program that implements in addition to the Fleiss' Kappa, several other multiple-

---

[1] Actually it is the book that Fleiss wrote in 1981 that was revised by others in 2003

rater agreement coefficients that were proposed in the literature.

I must say that, I did like very much the way **SAS** implemented the calculation of the Cicchetti-Allison, and Fleiss-Cohen weights for computing the weighted Kappa coefficient. What I mean specifically, is that if you define the categories as character-type values, then the weights are calculated based on sequential integer values from 1 to the number of categories. However, if your categories are numeric, then these numeric values are automatically used for defining the weights. This implementation gives the user considerable latitude for customizing the weights. Some researchers have contacted me about the availability of software products that would allow them to use their own set of weights. Although you can always use your weights, I do not recommend that approach because of the possible abuses that may result from such anarchy. Instead, I recommend that practitioners use numeric categories and change their values to reflect the way they view the seriousness of some disagreements. I will further discuss about this issue later in this book. While the **FREQ** procedure does not output the weights it has used[2], the solutions I propose always output these weights. I do believe that researchers should see the weights that were used for calculating weighted agreement coefficients.

1.3 Agreement Coefficient Options Limited to Kappa

Although several agreement coefficients have been proposed in the literature, **SAS** has only implemented **Kappa** in the **FREQ** procedure. I understand very well that the widespread use of Kappa must have played an important role in the decision that **SAS** developers made. But any researcher with some experience with the use of Kappa must know by now that there are situations in practice where **Kappa** will produce rather strange results. These

---

[2]Maybe I should say I haven't found a way to print them out from **SAS**

problems, largely known in the literature as the Kappa paradoxes have led several researchers to look for alternative coefficients. The Brennan-Prediger coefficient is one possible option that I like (see Brennan & Prediger (1981)). I also proposed a few years ago, the AC_1, which I considered to be a refinement of the Brennan-Prediger proposal (see Gwet (2008a)).

In the case of multiple raters (3 or more), researchers may want to explore alternative generalized coefficients due to Conger (1980), Fleiss (1971), Brennan-Prediger (1980), Gwet (2008a). SAS does not offer any of these options presently. I will present in chapters 5 and 6, a SAS macro that implements all of these agreement coefficients, and a few more, and will show you step by step how it can be used. From organizing input data to specifying the parameters, PC SAS, and SAS Enterprise Guide users will learn to compute various agreement coefficients and their associated precision measures within a short period of time.

1.4 The Diagonal Problem

The FREQ procedure of SAS was developed many years before the AGREE and KAPPA options were added to it. But the addition of these options did not change the broad way the FREQ procedure looks at a contingency table. To be concrete, let us consider the following contingency table:

Table 1.1: Distribution of 100 Patients by Rater and Level of Pain

		Rater 1			
		Moderate	No	Severe	Total
Rater 2	Mild	25	5	7	37
	No	6	24	4	34
	Severe	11	1	17	29
	Total	42	30	28	100

1.5. The Unbalanced-Table Problem.

At first sight, Table 1.1 resembles any ordinary frequency table showing the distribution of 100 patients by rater and pain level. But the raters had to score the patients on a 4-level scale. Each rater only used 3 of the 4 levels, with the 3 levels used not being the same. Although the information reported in Table 1.1 is accurate, the diagonal in this case does not represent agreement as we normally expect it in the area of inter-rater reliability. The **AGREE** and **KAPPA** options of the **FREQ** procedure however, will treat all diagonal elements of Table 1.1 as representing agreement, leading to wrong results. In other words, Table 1.1 is a contingency table, but it is not an agreement table. It is the agreement table that you need to compute inter-rater reliability adequately. The correct agreement table associated with Table 1.1 is the following:

Table 1.2: Distribution of 100 Patients by Rater and Level of Pain

		\multicolumn{4}{c}{Rater 1}	
		Mild	Moderate	No	Severe	Total
	Mild	0	25	5	7	37
	Moderate	0	0	0	0	0
Rater 2	No	0	6	24	4	34
	Severe	0	11	1	17	29
	Total	0	42	30	28	100

In Table 1.2, all diagonal elements now represent agreement. Therefore, when using the **FREQ** procedure of **SAS**, you need to ensure that the contingency table being used is in fact an agreement table. This issue is discussed in great details in chapters 2 and 3, and simple solutions are proposed.

1.5 The Unbalanced-Table Problem

Consider Table 1.3 representing the distribution of 100 elderly patients by rater and level of function. Once again, this table is a normal contingency table in a traditional sense. One thing however stands out, which is that rater 1 has used one function scale level more than rater 2. This situation led to the unbalanced table we have with 3 columns and only 2 rows. Therefore our contingency table does not have a diagonal. With no diagonal, the **FREQ** procedure of **SAS** cannot compute the Kappa coefficient nor any other related statistics.

Table 1.3: Distribution of 100 Patients by Rater & Functional Level

		Rater 1			
		Independent	Assistance	Dependent	**Total**
	Independent	25	5	7	37
Rater 2	Dependent	17	25	21	63
	Total	42	30	28	100

We have another case here where the contingency table is not an agreement table. With no diagonal in the table, **SAS** will simply ignore our request to compute Kappa. Most mainstream statistical techniques such as the Chi-square test that have historically been implemented in the **FREQ** procedure can work just fine with any contingency table. Moreover, the contingency tables typically analyzed in statistics involve 2 variables with different levels, and only the broad table configuration really matters. But the agreement table is a very special contingency table, which must be handled with care. I refer to this issue as the unbalanced-table issue, and further discuss it in chapters 2, and 3. I also propose simple solutions for resolving it.

The agreement table based on Table 1.3 data is given by Table 1.4. Now we have a balanced table with identical categories placed in the same order rowwise and columnwise. An agreement table ready for inter-rater reliability analysis.

1.6. The Ordinal Data Problem.

Table 1.4: Distribution of 100 Patients by Rater & Functional Level

		Rater 1			
		Independent	Assistance	Dependent	**Total**
	Independent	25	5	7	37
Rater 2	Assistance	0	0	0	0
	Dependent	17	25	21	63
	Total	42	30	28	100

1.6 Then Ordinal Data Problem

Several authors in the literature have advocated the use weighted agreement coefficients so that disagreements, which are more serious and those which are less serious do not receive the same treatment. It follows from table 1.4 that Rater 1 has considered dependent 7 patients who rater 2 considered independent. Here is a disagreement with obviously far more serious consequences that the one that occurred on the 5 patients raters 1 and 2 classified into the assistance and independent groups respectively. The more serious disagreements generally receive a smaller weight than the less serious ones.

By default, **SAS** sorts categories in alphabetical order for treatment. It means that Table 1.4 data for example will be organized as shown in Table 1.5 below.

Table 1.5: Distribution of 100 Patients by Rater & Functional Level

		Rater 1			
		Assistance	Dependent	Independent	**Total**
	Assistance	0	0	0	0
Rater 2	Dependent	25	21	17	63
	Independent	5	7	25	37
	Total	42	30	28	100

Both tables should a priori produce the same results. That is only true if you do not care about the weighted Kappa. If you do, then you will want to know that when computing the weighted Kappa from Table 1.5, **SAS** considers the Assistance-Independent disagreement to be more serious than the Dependent-Independent disagreement by assigning the smallest weight (typically a 0 weight) to the former. It is actually the Dependent-Independent disagreement that must be zero-weighted, as it is the one that should not be given any credit towards agreement. Therefore, when comes to agreement coefficient, even the way categories are labeled matters. I discuss the weighting further in chapter 4.

CHAPTER 2

Agreement Coefficients for 2 Raters: A Review

2.1 Overview

This chapter presents a non-mathematical introduction to a number of agreement coefficients that have been advocated in the literature when the number of raters is limited to 2. By opting for a non-mathematical presentation, I have focussed on getting the reader develop a quick understanding of how agreement coefficients are obtained through examples. Some readers may wonder, why I would want to show the steps for computing agreement coefficients, when it is **SAS** that will be doing the actual calculations. My answer is that, **SAS** will only do what you tell him to do. Unless you have a basic understanding of what you are doing, chances are **SAS** will give you wrong results very often. Secondly, a good way to ensure that you are doing things right is to start with a dataset simple and small enough to allow for manual calculations. Our manually-obtained results can then be compared with what **SAS** has produced. This can be done only if you understand the steps for calculating agreement coefficients yourself.

I will start in section 2.2 with the simplest reliability experiment possible, which involves 2 raters and 2 categories and will show step by step the procedures for computing various agreement coefficients. In section 2.3, I extend the discussion to the more general situation that involves 2 raters and 3 categories or more. Although, the situation involving 3 categories or more is not fundamentally different from the 2-category situation, it does introduce the new problem of different types of disagreements, some being more serious than others. It is this problem that led to the development of weighted agreement coefficients discussed in section 2.4. You will learn about the most widely-used weights suggested in the literature, and will learn how you may use your own custom weights.

2.2 Agreement for 2 Raters & 2 Categories

Ratings collected from an inter-rater reliability experiment involving 2 raters and a 2-category nominal scale are typically displayed in a 2 × 2 contingency table similar to Table 2.1. Table 2.1 shows the distribution of 39 patients by clinician, and relevance of lateral shift based on a low back pain assessment method applied by two clinicians 1 and 2 (see Kilpikoski et al. (2002)). Table 2.1 indicates that both clinicians agreed that lateral shift was relevant for 22 of the 39 patients, and also agreed that it was irrelevant on 11 patients. However, they also disagreed on the later shift relevance on 6 patients. The letters a, b, c, and d in the table are the abstract representations of the counts of patients.

Note that while the contingency table is a convenient and condensed way to display rating data, it also conceals patient-level information[1]. Luckily, this condensed information is all **SAS** needs to compute Kappa and its standard error. **SAS** however, requires the rating data to be organized as show in Table 2.2. Only 3 variables are necessary (Clinician1, Clinician2, and Count) and the number of records matches the total number of pairs of categories that the 2 raters can form - in this case that number is 4. *But there is one*

---

[1] For example the category into which the clinicians classified a specific patient.

2.2. Agreement for 2 Raters & 2 Categories.

problem. The FREQ procedure of SAS requires each rater to utilize all categories considered in the experiment. That is, if one clinician classifies all 39 subjects into the "Relevant" and none in the "Not Relevant", PROC FREQ will not work. We provide a solution to this problem in chapter 3.

Table 2.1: Distribution of n subjects by rater and response category

Clinician 1	Clinician 2		Total
	Relevant	Not relevant	
Relevant	a [22]	b [2]	$m1R$ [24]
Not Relevant	c [4]	d [11]	$m1N$ [15]
Total	$m2R$ [26]	$m2N$ [13]	n [39]

Table 2.2: Left Shift Relevance Ratings on 39 Patients by 2 Clinicians

Clinician 1	Clinician 2	Count
Relevant	Relevant	22
Relevant	Not Relevant	2
Not Relevant	Relevant	4
Not Relevant	Not Relevant	11

Kappa Coefficient

The general form of Kappa

$$\text{Kappa} = \frac{\text{Agreement Probability} - \text{Chance-Agreement Probability}}{1 - \text{Chance-Agreement Probability}},$$
$$= \frac{\text{AP} - \text{CAP}}{1 - \text{CAP}}.$$

The Agreement Probability (AP) is obtained by summing all diagonal elements of Table 2.1 and by dividing this sum by the total number of patients. It represents the relative number of times both clinicians agree on the classification of a patient. The Chance-Agreement Probability (CAP) on the other hand is solely based on the marginals of Table 2.1 and supposedly represents the propensity for both clinicians to agree by pure chance[2]. Both AP and CAP can be calculated using the notations of Table 2.1 as follows :

$$\text{AP} = (a+d)/n, \text{ and}$$
$$\text{CAP} = (m1R/n)(m2R/n) + (m1N/n)(m2N/n).$$

Using Table 2.1 numbers, these 2 probabilities are calculated as follows :

$$\text{AP} = \frac{22+11}{39} = 0.8462, \text{ and}$$
$$\text{CAP} = (24/39) \times (26/39) + (15/39) \times (13/39),$$
$$= 0.41026 + 0.12821 = 0.5385.$$

Consequently, the extent of agreement between the clinicians as quantified by Kappa is obtained as follows :

$$\text{Kappa} = \frac{0.8462 - 0.5385}{1 - 0.5385} = \frac{0.3077}{0.4615} = 0.6667.$$

Although the SAS system only implements Kappa, researchers often want to experiment with some alternative agreement coefficients that I will briefly describe here. These alternative statistics are implemented in the **AgreeStat_2SAS** SAS macro to be discussed in chapter 4.

---

[2]This interpretation of the meaning of CAP suggested by Cohen (1960) is as widely used as it is questionable. See Gwet (2010, in Chapter 2) for a detailed discussion

2.2. Agreement for 2 Raters & 2 Categories.

Scott's π-Coefficient

Scott (1955) proposed the Pi (or π) -agreement coefficient. Scott's coefficient has the same general form as Kappa, which depends on the 2 agreement probabilities AP, and CAP. The AP for both statistics is calculated the exact same way. The only difference between Kappa and Pi resides in the way CAP is calculated. The CAP associated with Scott's Pi is calculated (using Table 2.1 notations) as follows :

$$\begin{aligned}\text{CAP} &= \left(\frac{(m1R/n) + (m2R/n)}{2}\right)^2 + \left(\frac{(m1N/n) + (m2N/n)}{2}\right)^2, \\ &= \left(\frac{(24/39) + (26/39)}{2}\right)^2 + \left(\frac{(15/39) + (13/39)}{2}\right)^2, \\ &= 0.4109 + 0.1289 = 0.5398.\end{aligned}$$

Note that the two components (in parentheses) of this expression represent the propensities for the raters to categorize a subject into the "relevant" and "Not Relevant" categories respectively.

> Cohen (1960) who published his Kappa coefficient 5 years after Scott's Pi was proposed, pointed out that the CAP probability associated with Pi assumes that both raters categorize subjects with the same propensity. This argument was used to justify the formulation of CAP for Kappa.

While Cohen's argument may have some merit. Our investigation indicated that for all practical purposed, both formulations for the CAP lead to very similar results.

The CAP value calculated above leads to the following Scott π-statistic :

$$\text{Pi} = \frac{0.8462 - 0.5398}{1 - 0.5398} = 0.6658.$$

Bennet's S-Coefficient

Bennet et al. (1954) proposed an agreement coefficient referred to as the S-index, which is based on the same form as **Kappa** and Pi, and uses the same Agreement Probability (**AP**). However the Chance-Agreement Probability (**CAP**) associated with Bennet's S-index is different and defined as follows :

$$\mathsf{CAP} = 1/2 \quad \textit{(yes, as simple as that)}.$$

This **CAP** simply indicates that each rater will choose either category by chance with a probability of 1/2 (note that 2 here represents the number of categories under investigation). Brennan and Prediger (1981) extended this expression to the more general situation where the number of categories is arbitrary. If this case, chance-agreement probability is calculated as **CAP**=1/(**Number of Categories**). For 2 categories, Bennet's S-index is defined as follows :

$$\mathsf{S\text{-}Index} = 2 \times \mathsf{AP} - 1.$$

This simple coefficient appears to work reasonably well in most practical situations. Using Table 2.1 data, Bennet's S-index is calculated as follows :

$$\mathsf{S\text{-}Index} = 2 \times 0.8462 - 1 = 0.6924.$$

Gwet's AC_1-Coefficient

The inter-rater reliability literature abounds with examples where Kappa and many alternative coefficients do not work well (see Gwet (2010), or Cicchetti & Feinstein (1990) for a detailed discussion). To correct these problems Gwet (2008a) proposed the AC_1 coefficient, which shares the form as Kappa and uses the same **AP** probability.

2.2. Agreement for 2 Raters & 2 Categories.

However, the CAP associated with the AC_1 is defined (using Table 2.1's notations) as follows :

$$CAP = 2 \times P_R(1 - P_R) \text{ where } P_R = \frac{(m1R/n) + (m2R/n)}{2}.$$

The quantity P_R represents the propensity for classifying a patient into the "Relevant" category, and is quantified as follows :

$$P_R = \frac{(24/39) + (26/39)}{2} = (0.6154 + 0.6667)/2 = 0.6411.$$

This leads to a CAP value of $CAP = 2 \times 0.6411 \times (1 - 0.6411) = 0.4602$, and to an AC_1 value of,

$$AC_1 = \frac{0.8462 - 0.4602}{1 - 0.4602} = 0.7151.$$

AC_1 is the agreement coefficient that Gwet (2008a) recommends as a more robust alternative to the Kappa coefficient.

> *The inter-rater reliability literature often discusses the concept of "weights" in the context of weighted Kappa. Agreement coefficients can be weighted if the researcher considers certain types of disagreements to be more serious (in a sense) than others. The less serious disagreements are treated as partial agreements and receive a larger weight, while the more serious ones are treated as total disagreements and could receive a weight as small as 0.*
>
> *However, the weight concept is irrelevant with only 2 categories, since this situation only offers a single type of disagreement. For this reason, the SAS system does not produce the weighted analysis when the number of categories is limited to 2.*

2.3 Agreement for 2 Raters & 3 Categories or More

Quantifying the extent of agreement among two raters when the number of categories is 2 is not fundamentally different when the number of categories is 3 or more. The 4 agreement coefficients discussed in the previous section (i.e. Cohen's **Kappa**, Scott's Pi, Bennet's S-index, and Gwet's **AC**$_1$) can be extended in natural way. The main motivation for discussing the case of 3 categories or more in a separate section is the possibility to utilize the concept of weight with agreement coefficients for a more effective treatment of disagreements. Table 2.3 will be referred to throughout this section and beyond, and contains hypothetical data on the severity of movement-related pain at the shoulder joint, taken from 100 patients on 2 occasions. Pain severity is evaluated on a scale of 4 levels labeled as "No Pain", "Mild Pain", "Moderate Pain", and "Severe Pain." The 2 tests (Test 1 and Test 2) are the 2 raters whose agreement extent is to be quantified.

Table 2.3: Hypothetical Test-Retest Ratings of Movement-Related Pain at the Shoulder Joint

Test 1	Test 2				Total
	No Pain	Mild	Moderate	Severe	
No Pain	a[15]	b[3]	c[1]	d[1]	t1N[20]
Mild	e[4]	f[18]	g[3]	h[2]	t1MI[27]
Moderate	i[4]	j[5]	k[16]	l[4]	t1MO[29]
Severe	m[1]	n[2]	o[4]	p[17]	t1S[24]
Total	t2N[24]	t2MI[28]	t2MO[24]	t2S[24]	n [100]

For all 4 agreement coefficients, the agreement probability (AP) is still obtained by summing all diagonal elements and dividing the sum by the number of subjects, which is 100. That is,

$$\text{AP} = \frac{15+18+16+17}{100} = \frac{66}{100} = 0.66.$$

2.3. Agreement for 2 Raters & 3+ Categories.

▶ The chance-agreement probability (CAP) associated with Kappa is obtained by first multiplying the two table margins elementwise[3], by summing all 4 products, and by finally dividing the sum by the number of subjects n squared. Using actual numbers from Table 2.3 leads to the following CAP :

$$\text{CAP} = \frac{(20 \times 24) + (27 \times 28) + (29 \times 24) + (24 \times 24)}{100},$$
$$= 2508/10{,}000 = 0.2508.$$

Using the AP calculated above leads to the following Kappa coefficient :

$$\text{Kappa} = \frac{0.66 - 0.2508}{1 - 0.2508} = 0.5462.$$

▶ The CAP for Scott's Pi coefficient is obtained by first computing the classification probabilities associated with the 4 categories of Table 2.3. For the "No Pain" category for example, the classification probability is calculated as $(20/100+24/100)/2 = 0.44/2 = 0.11$. Next, each of the 4 probabilities will be squared, and finally the sum of all squared values will yield CAP. The classification probabilities associated with the "Mild", "Moderate", and "Severe" categories are respectively given by $(27/100 + 28/100)/2 = 0.275$, $(29/100 + 24/100)/2 = 0.265$, and $(24/100 + 24/100)/2 = 0.24$. Thus,

$$\text{CAP} = 0.11^2 + 0.275^2 + 0.265^2 + 0.24^2 = 0.2156.$$

The same AP value used to obtain Kappa is used again to compute Scott's Pi as follows :

$$\text{Pi} = \frac{\text{AP} - \text{CAP}}{1 - \text{CAP}} = \frac{0.66 - 0.2156}{1 - 0.2156} = 0.5666.$$

---
[3]This operation will produce 4 products $t1N \times t2N$, $t1MI \times t2MI$, $t1MO \times t2MO$, and $t1S \times t2S$

▶ Brennet's S-index was extended to 3 categories or more by Brennan & Prediger (1981) to become the Brennan-Prediger (BP) coefficient. It utilizes the same AP as the other coefficients, and uses a CAP value of 1/4 (4 being the number of categories)[4]. Therefore, BP is obtained as follows :

$$BP = \frac{AP - CAP}{1 - CAP} = \frac{0.66 - 0.25}{1 - 0.25} = 0.5467.$$

▶ To obtain the CAP associated with the AC_1, we will use the classification probabilities calculated for Scott's Pi coefficient. These probabilities are 0.11, 0.275, 0.265, and 0.24 for the "No Pain", "Mild Pain", "Moderate Pain", and "Severe Pain" respectively. Chance-agreement probability is obtained as follows :

$$CAP = \frac{1}{4-1}\Big[0.11 \times (1-0.11) + 0.275 \times (1-0.275) +$$
$$0.265 \times (1-0.265) + 0.24 \times (1-0.24)\Big],$$
$$= 0.2248.$$

where 4 in the denominator represents the number of categories. The AC_1 coefficient is then obtained as follows :

$$AC_1 = \frac{AP - CAP}{1 - CAP} = \frac{0.66 - 0.2248}{1 - 0.2248} = 0.5614.$$

---

[4]This means that CAP=1/4=0.25

2.4 Weighting Agreement Coefficients

As mentioned in the previous section, the only fundamental difference between 2-level and 3-level (or more) nominal scales is the occasional need to weight agreement coefficients. Why would you need to weight? To understand the motivation behind weighting, consider again the 4 categories ("No Pain", "Mild Pain", "Moderate Pain", "Severe Pain") used in Table 2.3. When 2 raters disagree about the categorization of a patient, it may be because the first rater evaluated the pain as being mild, while the second deemed it moderate. Although a moderate pain is supposedly a little more acute than a mild pain, this disagreement does not appear to be significant. However, if the first and second raters categorize a pain as "No Pain" and "Severe Pain" respectively, then we know there is a serious disagreement issue among raters that must be addressed. A "No-Pain" pain is supposed to provide no discomfort at all to the subject, while a severe pain would provide maximum discomfort. It is the need to take into consideration the seriousness of the disagreement in the calculation of agreement coefficients that led the experts to introduce the notion of weighting.

In a nutshell, the weighting process comes down to downweighting[5] serious disagreements so as to dampen their impact of the agreement coefficient, while upweighting[6] less serious disagreements (also referred to as partial agreements in Cohen's terminology) in order to make their impact bigger. The weighting is implemented in practice by assigning a number (typically between 0 and 1) to each pair of categories that may be used by the two raters. For example both tests (or raters) in Table 2.3 may use No-No, Mild-Mild, Moderate-Moderate, or Severe-Severe each time they agree. The outcome of a subject categorization could also be Mild-Severe

---

[5] Downweighting amounts to assigning a smaller weight
[6] Upweighting amounts to assigning a bigger weight

for example when the raters disagree. All pairs with identical categories are often assigned a weight of 1 to give then full (agreement) credit. Pairs where the 2 categories are different such as Mild-Moderate will be assigned a weight value smaller than 1 that decreases with the severity of the disagreement. It is even common in the literature to see pairs made up of the 2 most extreme categories (e.g. No-Severe) have a weight of 0.

The magnitude of a weight is an attribute of a specific pair of categories. The weight's magnitude characterizes the seriousness of the disagreement described by the 2 categories that make up the pair, and does not depend on the raters nor on the subjects. Although a number of different types of weights have been proposed in the literature, the SAS system has implemented only two of them called the Linear Weights (also referred to in SAS as the Cicchetti-Allison (CA) weights[7]), and the Quadratic Weights (also referred to in SAS as the Fleiss-Cohen (FC) weights[8]).

The Linear Weights (or CA Weights)

▶ The Cichetti-Allison weights (or CA-weights) associated with Table 2.3 are shown in Table 2.4. The diagonal cells representing agreement are assigned a weight of 1, while an off-diagonal cell has a weight that decreases with the seriousness of the disagreement.

▶ The CA-weights are obtained by first assigning the sequential numbers 1, 2, 3, 4 to the 4 categories under investigation (i.e. 1 is associated with "No Pain" while 4 is associated with "Severe Pain"). For two categories such as 2 and 4, the weight

---

[7] These weights were suggested by Cicchetti & Allison (1971)
[8] These weights were suggested by Fleiss & Cohen (1973)

2.4. Weighting: 2 Raters & 3+ Categories.

is calculated as follows :

$$\text{CA-Weight} = 1 - \frac{\text{Absolute Value}(2-4)}{4-1} = 1 - \frac{2}{3} = 0.3333,$$

where the denominator (4-1) represents the number of categories minus 1.

Table 2.4: Linear Weights

	No Pain	Mild	Moderate	Severe
No Pain	1.0	0.667	0.333	0.0
Mild	0.667	1.0	0.667	0.333
Moderate	0.333	0.667	1.0	0.667
Severe	0.0	0.333	0.667	1.0

The Quadratic Weights (or FC Weights)

▶ The Fleiss-Cohen weights (or FC-weights) associated with Table 2.3 are shown in Table 2.5. Diagonal cells representing agreement are again assigned a weight of 1, while off-diagonal cells are assigned weights that decrease with the seriousness of the disagreement.

▶ FC-weights are obtained by first assigning the sequential numbers 1, 2, 3, 4 to the 4 categories under investigation (i.e. 1 is associated with "No Pain" while 4 is associated with "Severe Pain"). For two categories such as 2 and 4, the weight is calculated as follows :

$$\text{FC-Weight} = 1 - \frac{(2-4)^2}{(4-1)^2} = 1 - \frac{(-2)^2}{3^2} = 1 - \frac{4}{9} = 0.556,$$

where the denominator (4-1) represents the number of categories minus 1.

Table 2.5: Quadratic Weights

	No Pain	Mild	Moderate	Severe
No Pain	1.0	0.889	0.556	0.0
Mild	0.889	1.0	0.889	0.556
Moderate	0.556	0.889	1.0	0.889
Severe	0.0	0.556	0.889	1.0

Calculating Weighted Coefficients

The purpose of this sub-section is two-fold:

a) To use an example and briefly show you what **SAS** does behind the scene when calculating the weighted Kappa coefficient,

b) To show you how agreement coefficients other than Kappa can be weighted. I have implemented these alternative coefficients in a SAS macro to be discussed in the next chapter[9].

All weighted agreement coefficients still have the same general form as their unweighted versions. They are expressed as a ratio of the difference between the weighted agreement probability (**WAP**) and the weighted chance-agreement probability (**WCAP**) to the probability of no agreement by chance[10]. For example the weighted Kappa, which we will denote as **WKAPPA** takes the following form :

$$\text{WKAPPA} = \frac{\text{WAP} - \text{WCAP}}{1 - \text{WCAP}}..$$

While all 4 agreement coefficients under investigation here share the same **WCAP**, each coefficient will have a distinct **WCAP**.

---

[9]The SAS system has implemented Kappa as the only agreement statistic. Therefore, a macro was necessary to implement alternatives

[10]This probability is obtained as 1-WCAP

2.4. Weighting: 2 Raters & 3+ Categories.

Calculating WAP

[1] Using Table 2.3, create a new table containing the relative numbers (i.e. proportions) of subjects by rater and category. The new table will look like Table 2.6. *It follows from Table 2.6 that 3% of subjects were categorized as Mild and Moderate according to Test 1 and Test 2 respectively.*

[2] Choose a set of weights to use (either CA or FC weights). For illustration purposes, let us consider FC weights (see Table 2.5).

[3] Multiply both tables (2.5 and 2.6) elementwise to produce Table 2.7.

[4] Sum all Table 2.7 elements to obtain the **WAP** (in this case, WAP=0.91451).

Table 2.6: Hypothetical Test-Retest Ratings of Movement-Related Pain at the Shoulder Joint

Test 1	Test 2				Total
	No Pain	Mild	Moderate	Severe	
No Pain	a[0.15]	b[0.03]	c[0.01]	d[0.01]	t1N[0.20]
Mild	e[0.04]	f[0.18]	g[0.03]	h[0.02]	t1MI[0.27]
Moderate	i[0.04]	j[0.05]	k[0.16]	l[0.04]	t1MO[0.29]
Severe	m[0.01]	n[0.02]	o[0.04]	p[0.17]	t1S[0.24]
Total	t2N[0.24]	t2MI[0.28]	t2MO[0.24]	t2S[24]	n [1.00]

Table 2.7: Elementwise Multiplication of Tables 2.5 and 2.6

Test 1	Test 2			
	No Pain	Mild	Moderate	Severe
No Pain	0.15	0.02667	0.00556	0
Mild	0.03556	0.18	0.02667	0.01112
Moderate	0.02224	0.04445	0.16	0.03556
Severe	0	0.01112	0.03556	0.17

Calculating Kappa's WCAP

> [1] Obtain the "direct" product of the Table 2.6 marginals (Test1 marginals are in the rightmost column, and Test2 marginals in the bottom row) by multiplying then elementwise as shown in Table 2.8.
>
> [2] Multiply the weight table (Table 2.5) and Table 2.8 elementwise to produce Table 2.9.
>
> [3] Sum all Table 2.9 elements to obtain Kappa's WCAP (in this case, WCAP=0.7398).

Therefore, the weighted Kappa coefficient is calculated as follows:

$$\text{WKappa} = \frac{\text{WAP} - \text{WCAP}}{1 - \text{WCAP}} = \frac{0.9145 - 0.7398}{1 - 0.7398} = 0.6714.$$

Table 2.8: Elementwise Product of Table 2.6 Marginal Probabilities

Test1	Test2			
	0.24	0.28	0.24	0.24
0.20	0.0480	0.0560	0.048	0.0480
0.27	0.0648	0.0756	0.0648	0.0648
0.29	0.0696	0.0812	0.0696	0.0696
0.24	0.0576	0.0672	0.0576	0.0576

2.4. Weighting: 2 Raters & 3+ Categories.   - 25 -

Table 2.9: Elementwise Product of Tables 2.5 and 2.8[a]

Test1	Test2			
	0.24	0.28	0.24	0.24
0.20	0.0480	0.0498	0.0267	0
0.27	0.0576	0.0756	0.0576	0.0360
0.29	0.0387	0.0722	0.0696	0.0619
0.24	0	0.0374	0.0512	0.0576

[a]Note that Table 2.9 is a weighted version of Table 2.8

Calculating Scott-Pi's WCAP

[1] Using Table 2.6, obtain the classification probability of each category, by averaging the 2 marginals associated with that category (e.g. $(0.24 + 0.20)/2 = 0.22$ is the no-pain classiciation probability). The 4 probabilities of interest are 0.22, $(0.28 + 0.27)/2 = 0.275$, $(0.24 + 0.29)/2 = 0.265$, and $(0.24 + 0.24)/2 = 0.24$.

[2] Obtain the "direct" product of the classification probabilities {0.22, 0.275, 0.265, 0.24} as shown in Table 2.10

[3] Multiply the weight table (Table 2.5) and Table 2.10 elementwise to produce Table 2.11.

[4] Sum all Table 2.11 elements to obtain Pi's WCAP (in this case, WCAP=0.7403.

Therefore, the weighted Scott's Pi coefficient is calculated as follows:

$$\text{WPi} = \frac{\text{WAP} - \text{WCAP}}{1 - \text{WCAP}} = \frac{0.9145 - 0.7403}{1 - 0.7403} = 0.6708.$$

Table 2.10: Elementwise Product of the Classification Probabilities {0.22, 0.275, 0.265, 0.240}

Test1	Test2			
	0.220	0.275	0.265	0.24
0.220	0.0484	0.0605	0.0583	0.0528
0.275	0.0605	0.0756	0.0729	0.066
0.265	0.0583	0.0729	0.0702	0.0636
0.240	0.0528	0.0660	0.0636	0.0576

Table 2.11: Elementwise Product of Tables 2.5 and 2.10[a]

Test1	Test2			
	0.220	0.275	0.265	0.240
0.20	0.0484	0.0538	0.0324	0
0.27	0.0538	0.0756	0.0648	0.0367
0.29	0.0324	0.0648	0.0702	0.0565
0.24	0	0.0367	0.0565	0.0576

[a] Note that Table 2.11 is a weighted version of Table 2.10

Calculating Brennan-Prediger (BP)'s WCAP

[1] Sum all elements of the weight table. For FC-weights of Table 2.5, this sum is 11.558.

[2] Compute WCAP = 11.558/16 = 0.7224 (note : 16 = 4 × 4, where 4 is the number of categories, and 11.558 the sum of step 1).

Therefore, the weighted BP coefficient is calculated as follows :

$$\text{W-BP} = \frac{\text{WAP} - \text{WCAP}}{1 - \text{WCAP}} = \frac{0.9145 - 0.7224}{1 - 0.7224} = 0.6920.$$

2.4. Weighting: 2 Raters & 3+ Categories. - 27 -

Calculating Gwet's AC$_1$'s WCAP

> [1] Multiply each element of Table 2.10 by the number of categories (i.e. 4), and subtract this product from 1. This leads to Table 2.12.
>
> [2] Multiply the weight table (Table 2.5) and Table 2.12 element-wise to produce Table 2.13.
>
> [3] Sum all Table 2.13 elements and divide the sum by $4 \times (4-1)$ where 4 is the number of categories to obtain AC$_1$'s WCAP (in this case, WCAP $= 8.5968/(4 \times 3) = 8.5968/12 = 0.7164$).

Therefore, the weighted AC$_1$ coefficient is calculated as follows:

$$\text{WAC}_1 = \frac{\text{WAP} - \text{WCAP}}{1 - \text{WCAP}} = \frac{0.9145 - 0.7164}{1 - 0.7164} = 0.6985.$$

Table 2.12: $1 - 4 \times$ Table 2.10 Elements

Test1	Test2			
	0.220	0.275	0.265	0.24
0.220	0.8064	0.7580	0.7668	0.7888
0.275	0.7580	0.6975	0.7085	0.736
0.265	0.7668	0.7085	0.7191	0.7456
0.240	0.7888	0.7360	0.7456	0.7696

Table 2.13: Elementwise Product of Tables 2.5 and 2.12[a]

Test1	Test2			
	0.220	0.275	0.265	0.240
0.20	0.8064	0.6739	0.4263	0
0.27	0.6739	0.6975	0.6299	0.4092
0.29	0.4263	0.6299	0.7191	0.6628
0.24	0	0.4092	0.6628	0.7696

[a]Note that Table 2.13 is a weighted version of Table 2.12

CHAPTER 3

Kappa and the FREQ Procedure of SAS

3.1 Overview

SAS implements the computation of Cohen's Kappa and weighted Kappa statistics as an option in its **FREQ** procedure. However, the number of raters to be analyzed must be limited to 2. The more general situation involving 3 raters or more cannot be analyzed using the **FREQ** procedure. That is, the many extensions of Kappa proposed by Fleiss (1971), Light (1971), Conger (1980) or Gwet (2008a) are not implemented in **SAS** 9.2 nor in any previous version. However, some **SAS** macros to be discussed in chapter 4 have been developed and can be used to compute these multiple-rater agreement coefficients.

To evaluate the extent of agreement among 3 raters or more, some researchers would average all pairwise Kappa coefficients obtained from all pairs of raters with the **FREQ** procedure. Although this approach is technically feasible and appears to be reasonable at first sight, it is actually flawed. Averaging for example the mean annual income of the female population and that of the male population will certainly not yield the average income of the whole population of males and females combined. What is needed is a multivariate approach that can quantify the extent of agreement among multiple raters as proposed by the authors mentioned above. To allow researchers to experiment with agreement coefficients other than Kappa, I developed a **SAS** macro that implements Scott's **PI**, Brennan-Prediger (1981) coefficient, Gwet's **AC$_1$** and others.

3.2 Organizing Your Data

The input file needed to compute Kappa with the **FREQ** procedure can take 2 forms, which are the Contingency Table and the Raw Scores.

(1) **The Contingency Table**

The researcher may have the distribution of subjects by rater and by category in the following form:

Rater 1	Rater 2		
	A	B	C
A	5	1	0
B	0	3	2
C	1	1	2

In this case the following **SAS** program will compute the Kappa as well as other statistics:

```
01   DATA RatingFile ;
02        INPUT Rater1$ Rater2$ Count;
03        DATALINES;
04   A A 5
05   A B 1
06   A C 0
07   B A 0
08   B B 3
09   B C 2
10   C A 1
11   C B 1
```

3.2 Organizing Your Data.

```
12    C C 2
13    ;
14    PROC FREQ DATA = RatingFile;
15        WEIGHT Count;
16        TABLES Rater1*Rater2/AGREE;
17        RUN;
```

Program 3.1: Basic SAS Program for Computing Kappa based on a Contingency Table.

(2) Raw Scores

Rating data collected from an inter-rater experiment do not normally present themselves in the form of a contingency table. Instead, that data is often in the form of a list of subjects, and the specific categories into which each rater classified them. Here is an example of such a configuration:

Subject	Rater 1	Rater 2
1	A	A
2	A	A
3	A	A
4	A	A
5	A	A
6	B	B
7	B	B
8	B	B
9	C	C
10	C	C
11	C	A
12	C	B
13	B	C
14	B	C
15	A	B

If the raw scores are all we have, there is no need to create a contingency table showing the distribution of subjects by rater and category to compute Kappa with the **FREQ** procedure. The raw scores can be supplied to **SAS** as shown in **Program 3.2**. Both Programs 3.1 and 3.2 will produce the same **Output 3.1**. This output is divided into the following three parts :

▶ The contingency table,

▶ The marginal homogeneity test,

▶ The Simple and Weighted Kappa Statistics

The Contingency Table

The first part of the output is a frequency table showing the distribution of subjects by rater and response category.

Practitioners should ALWAYS verify that diagonal cells define subjects that both raters classified into the exact same category. Actually, the FREQ procedure of SAS may well produce the following contingency table:

Rater 1	Rater 2			Total
	B	C	E	
A	[5]	1	0	6
B	0	[3]	2	5
C	1	1	[2]	4
Total	6	5	4	15

A priori there is nothing wrong with this table, which shows accurate counts, except that the circled diagonal elements do not represent agreement. The first diagonal cell for example indicates that raters 1 and 2 both classified 5 subjects into categories A and B respectively. This will lead to the wrong Kappa coefficient. This the "diagonal issue" previously mentioned. I will show later how to deal with it.

3.2 Organizing Your Data.

The Marginal Homogeneity Test

▶ The second part of the output shows the results of the marginal homogeneity testing based on the Bowker's test statistic. The most important statistic to look at in this table is the p-value. That is, Pr>S = 0.5062. When this number is smaller than 0.05, we can conclude that the marginals are not homogeneous (i.e. the hypothesis of homogeneity of marginals is rejected). In this example the homogeneity hypothesis is not rejected.

▶ Some authors (e.g. Zwick (1988)) have indicated that Kappa does not work well when the marginals are not homogeneous, hence the interest of this test. These authors recommend to first test for marginal homogeneity, and use Kappa only when the homogeneity hypothesis is not rejected as is the case in this example. This approach however does not have a solid statistical foundation and does not suggest an alternative coefficient to use. I believe that Kappa is what it is regardless of the homogeneity of the marginals. Alternative agreement coefficients could be considered when a practitioner is dissatisfied with Kappa (see Gwet (2008a)). I find the usefulness of this test questionable as far as rater agreement is concerned.

The Kappa Statistics

▶ The last and more interesting part of the output provides the "Simple Kappa" and "Weighted Kappa[1]" estimates, the associated ASE (Asymptotic Standard Errors) and the lower and upper bounds of the 95% confidence interval. Note that the term **Asymptotic** associated with "Standard Errors" is used to remind us that the standard error value calculated by **SAS** is based upon an equa-

---

[1]Note that when the number of categories is limited to 2, **SAS** does not produce weighted kappa statistics, because both simple and weighted kappa statistics are always identical in this case.

tion that is valid only when the number of subjects is "sufficiently" large. In practice, a group of 20 subjects may be considered sufficiently large to ensure the ASE validity, although the minimum could be lower if Kappa is high.

▶ The ASE itself is not necessarily very helpful in practice. However, the 95% confidence limits determine the range of values that we claim with 95% certainty, contain the "true" Kappa value. Note that the kappa value of 0.4932 is calculated based on a small group (or sample) of 15 subjects only. If we had used a larger group of 25 subjects for example, we would have obtained a different value for kappa. Such differences show 2 things:

(a) That Kappa is subject to what is known as sampling error, which is estimated by the ASE.

(b) That the Kappa value of 0.4932 produced by SAS is not the "true" value of the extent of agreement among raters 1 and 2. Instead, it merely is a rough approximation based on our specific sample of 15 subjects. Unfortunately, the true Kappa is unknown to us. How can we get it? Well, we will need to have the whole universe of subjects that the researcher is interested in. We are entering the domain of statistical inference here. Readers interested in a more elaborate discussion on these issues as they relate to agreement coefficients can read chapter 5 of Gwet (2010).

3.2 Organizing Your Data.

```
01  DATA RatingFile ;
02      INPUT Rater1$ Rater2$ ;
03      DATALINES;
04  A A
05  A A
06  A A
07  A A
08  A A
09  B B
10  B B
11  B B
12  C C
13  C C
14  C A
15  C B
16  B C
17  B C
18  A B
19  ;
20  PROC FREQ DATA = RatingFile;
21      TABLES Rater1*Rater2/AGREE;
22      RUN;
```

Program 3.2: Basic SAS Program for Computing Kappa based Raw Scores.

Chapter 3. *Kappa with SAS' FREQ Procedure*

The FREQ Procedure
Table of Rater1 by Rater2

Rater1   Rater2

Frequency Percent Row Pct Col Pct	A	B	C	Total
A	5 33.33 83.33 83.33	1 6.67 16.67 20.00	0 0.00 0.00 0.00	6 40.00
B	0 0.00 0.00 0.00	3 20.00 60.00 60.00	2 13.33 40.00 50.00	5 33.33
C	1 6.67 25.00 16.67	1 6.67 25.00 20.00	2 13.33 50.00 50.00	4 26.67
Total	6 40.00	5 33.33	4 26.67	15 100.00

Statistics for Table of Rater1 by Rater2

Test of Symmetry

Statistic (S)	2.3333
DF	3
Pr > S	0.5062

Kappa Statistics

Statistic	Value	ASE	95% Confidence Limits
Simple Kappa	0.4932	0.1797	0.1411 0.8454
Weighted Kappa	0.5408	0.1779	0.1920 0.8896

Sample Size = 15

Output 3.1. Output of SAS Programs 1 and 2.

3.3 Potential Data Problems

In the introductory chapter, I mentioned 2 potential serious problems that may be encountered when using the **FREQ** procedure of **SAS** to compute the Kappa coefficient. These are the unbalanced table problem and the diagonal problem. Both problems stem from the misinterpretation of certain types of rating data by the **FREQ** procedure. In this section, I will review both problems and will propose a possible solution to each.

The Unbalanced Table Problem

It is legitimate in a reliability experiment for one rater not to classify any subject into a particular category, while the other rater uses all available categories. This situation will lead to an unbalanced table. Let us consider the following frequency table as an example:

Table 3.1: Unbalanced Table

Rater 1	Rater 2 A	Rater 2 B
A	15	1
B	2	10
C	3	4

Rater 2 did not categorize any subject in group **C**. She used only 2 categories while Rater 1 used all 3. This difference in the number of categories used by both raters has resulted in an unbalanced frequency table. The **FREQ** procedure in this situation, will produce no kappa statistics - simple nor weighted. This is due to the fact that the **FREQ** procedure, which was developed decades before the kappa option was added to it, always deletes rows and columns, which contain only zeros.

The Diagonal Problem

Practitioners may also expect in a reliability experiment, a situation where both raters use the exact same number of categories of different types. For example, Rater 1 may use categories A, B, and C, when Rater 2 used categories A, B, and D in a reliability experiment where all 4 categories A, B, C, and D are the possibilities. This will lead to a perfectly balanced table, except that SAS will match A to A, B to B, and C to D. That is SAS will interpret the categorization of a subject in C and D by raters 1 and 2 as an agreement. This will necessarily lead to wrong results. Let us consider the following frequency table :

Table 3.2: Balanced Table with Problem Diagonal

Rater 1	Rater 2 A	B	D
A	15	1	0
B	2	10	2
C	3	4	7

If the above data are captured in SAS by only providing the information displayed in Table 3.2, wrong Kappa and weighted Kappa statistics will be produced. The "diagonal problem" is more serious than the unbalanced table problem since the former may go undetected, while the later will not.

Solutions to the Unbalanced-Table and Diagonal Problems

Although the unbalanced-table and diagonal problems are different in nature, any "good" solution to one of the problems will be good to the other. Therefore, I am going to recommend a single solution to both problems. Several authors have proposed a variety of ways to solve these two problems.

3.3 Potential Data Problems. - 39 -

▶ One method often suggested in the literature is to add fictitious (or dummy) observations to the input file in such a way that there is at least one subject in each category for each rater. The fictitious data will then be assigned a very small weight so as to reduce their impact on the kappa statistics. More on this approach can be found in Stein at al. (2005), or in SAS Usage Note 22883 at:

http://support.sas.com/kb/22/883.html

▶ A second approach, discussed by Liu and Hays (1999) and implemented in their SAS macro called "%kappa", consists of creating the correct (balanced) two-way contingency table from the input data set prior to computing the kappa statistics. The Liu-Hayes solution amounts to filling empty cells with zeros before computing the kappa statistics. The SAS macro that Liu and Hayes have written to implement their method can be downloaded at:

http://www.gim.med.ucla.edu/FacultyPages
/Hays/UTILS/WKAPPA.txt

Their paper describing the approach they used can also be downloaded at:

http://www2.sas.com/proceedings/sugi24/Stats/p280-24.pdf

I believe that these two approaches are more complex than necessary. The simplest and cleanest approach I have found is due to Crewson (2001). One may download the PDF file of his paper at:

http://www2.sas.com/proceedings/sugi26/p194-26.pdf

It consists of supplementing the initial rating dataset with additional fictitious ratings chosen so as to balance the frequency table, and to allow for the calculation of Kappa. The actual and fictitious ratings will be flagged using a variable that I decided to name **Flag**. The **Flag** variable will take a value of 1 for the actual ratings, and a missing value for the fictitious ones.

Using Table 3.1 data as example, and Crewson's approach, I like to show step by step how the Unbalanced Table problem can be resolved:

Step 1. Flagging Initial Rating Data

Read rating data in the SAS data step, create the Flag variable, and assign 1 to it. This data step could be written as follows:

```
01  DATA RatingFile;
02      INPUT Rater1$ Rater2$ Count;
03      Flag = 1;
04      DATALINES;
05  A A 15
06  A B 1
07  B A 2
08  B B 10
09  C A 3
10  C B 4
11  ;
```

Step 2. Creating Fictitious Ratings to Balance the Data

Create a SAS file (I call it DummyFile), which contains the fictitious ratings. This step is implemented is SAS as follows:

```
01  DATA DummyFile;
02      INPUT Rater1$ Rater2$;
03      Count = 1;
04      DATALINES;
05  A A
06  A B
07  A C
08  B A
09  B B
10  B C
11  C A
12  C B
13  C C
14  ;
```

3.3 Potential Data Problems.   - 41 -

Step 3. Adding Dummy Ratings to Actual Ratings

Combine fictitious and actual ratings into a single dataset that I named "NewRatingFile" (see Table 3.3 for its content). This task is implemented in SAS by concatenating the 2 datasets RatingFile and DummyFile as follows:

```
01  DATA NewRatingFile;
02       SET RatingFile DummyFile;
```

Step 4. Computing the Kappa Coefficient

To compute Kappa and weighted Kappa, use the FREQ procedure as follows:

```
01  PROC FREQ;
02       TABLES Flag*Rater1*Rater2/KAPPA NOPERCENT
              NOROW NOCOL;
03  RUN;
```

This FREQ procedure will compute the kappa statistics (weighted and unweighted) separately for actual ratings (Flag=1), and for fictitious ratings (Flag=.). The results are shown in Output 3.2.

Table 3.3: Content of Dataset NewRatingFile

Obs	Rater1	Rater2	Count	Flag
1	A	A	15	1
2	A	B	1	1
3	B	A	2	1
4	B	B	10	1
5	C	A	3	1
6	C	B	4	1
7	A	A	1	.
8	A	B	1	.
9	A	C	1	.
10	B	A	1	.
11	B	B	1	.
12	B	C	1	.
13	C	A	1	.
14	C	B	1	.
15	C	C	1	.

▶ In step 2, our dummy file contains all possible pairs of ratings that both raters can produce. However, only the rating pairs missing from the initial file are really necessary to balance the table. The decision to include all the possibilities is a safer alternative to the error-prone task of identifying all missing pairs.

▶ Only the top portion of Output 3.2 that is related to actual ratings (i.e. everything with the subtitle "Controlling for Flag=1") is of interest. Everything else should be ignored.

▶ Using a third variable in the TABLE statement as in step 4 with the variable Flag, can help analyze Kappa separately for different groups of subjects[2] and to compute an overall Kappa. The overall analysis of Kappa is based on the work Fleiss (see Fleiss et al (2003), chapter 18). *I personally have some reservations about the statistical foundation of the methodology these authors recommend for studying the overall Kappa.* I will further expand on this issue in another edition of the book.

▶ The solution I just presented resolves both the diagonal and the unbalanced-table problems. It is up to you to determine whether there is a need to create a "true" agreement table or not. Alternatively you may want to use the **SAS** macro I will present in chapter 4 to avoid having to verify the adequacy of your input data.

---

[2]These groups of subjects are also known as strata in **SAS** terminology

3.3 Potential Data Problems.

Output 3.2. Results of the Analysis of Table 3.1

```
                  The FREQ Procedure

              Table 1 of Rater1 by Rater2
                 Controlling for Flag=1
     Rater1      Rater2
```

Frequency	A	B	C	Total
A	15	1	0	16
B	2	10	0	12
C	3	4	0	7
Total	20	15	0	35

```
       Statistics for Table 1 of Rater1 by Rater2
                 Controlling for Flag=1
```

Test of Symmetry	
Statistic (S)	7.3333
DF	3
Pr > S	0.0620

Kappa Statistics

Statistic	Value	ASE	95% Confidence Limits	
Simple Kappa	0.5172	0.1076	0.3063	0.7281
Weighted Kappa	0.4740	0.1099	0.2586	0.6893

Sample Size = 35

The FREQ Procedure

Summary Statistics for Rater1 by Rater2
Controlling for Flag

Overall Kappa Coefficients

Statistic	Value	ASE	95% Confidence Limits
Simple Kappa	0.5172	0.1076	0.3063 0.7281
Weighted Kappa	0.4740	0.1099	0.2586 0.6893

Test for Equal Kappa Coefficients

Statistic	Chi-Square	DF	Pr > ChSq
Simple Kappa	0.0000	0	.
Weighted Kappa	0.0000	0	.

Effective Sample Size = 35
Frequency Missing = 9
WARNING: 20% of the data are missing

CHAPTER 4

Weighted Kappa & the FREQ Procedure of SAS

4.1 Overview

In chapter 3, we have learned that the **FREQ** Procedure always produces the simple and weighted each time one of the options **AGREE** or **KAPPA** is used with the **TABLE** statement (see Program 3.1 for example). Therefore, there is a priori nothing special that we need to do to obtain the weighted kappa. However, SAS offers 2 choices of weights to use for computing the weighted Kappa, which are the linear weights - also called the Cicchetti-Allison (CA) weights - and the Quadratic Weights - also known as the Fleiss-Cohen (FC) weights. I will discuss in section 4.2 how to select them, and how to define them properly.

In this chapter, I will also present a SAS macro program called **AgreeStat_2SAS** that I developed to compute various simple and weighted agreement coefficients including, Kappa, Brennan-Prediger, Gwet's AC_1, and Scott's Pi. This macro resolves the unbalanced table and the diagonal problems discussed in chapter 3, in addition to allowing the user to supply custom weights. Given the arbitrary nature of the FC and CA weights, being able to use custom weights is essential to some researchers.

4.2 The Weights

Program 3.1 in chapter 3 produces the following kappa-related statistics:

```
                   Kappa Statistics
Statistic         Value      ASE    95% Confidence Limits
Simple Kappa      0.4932   0.1797   0.1411        0.8454
Weighted Kappa    0.5408   0.1779   0.1920        0.8896
                    Sample Size = 15
```

There is a possible confusion in that SAS program between the term "Weighted Kappa" and the WEIGHT statement that appears in the FREQ procedure. In fact, the 2 are not related at all. Both Simple and Weighted Kappa use the Count variable the same way. What distinguishes the Simple (unweighted) and weighted Kappa is the use the Cicchetti-Allison (CA) weights (the default weights in SAS) in the calculation of the weighted kappa. If the user prefers the FC weights instead, Program 3.1 could be replaced with Program 4.1. The only difference between these 2 programs is the use of the option WT=FC with the TABLE statement. When no WT option is used, SAS will use CA weights by default as if the option WT=CA was used.

Note that a weight is not assigned to each subject being rated. Instead, a weight is assigned to each combination of categories that the 2 raters may choose. If A, B, and C are the 3 categories a subject can be classified into, then each of the pairs (A,A), (A,B), (A, C), (B, A), (B,B), (B,C), (C,A), (C,B), and (C,C), must be assigned a weight value, that typically varies from 0 to 1 depending on the "seriousness" of the disagreement. The more serious the disagreement, the lower the weight.

Now, let us see what the CA and FC weights look like :

4.2 The Weights.

```
01   DATA RatingFile;
02        INPUT Rater1$ Rater2$ Count;
03        DATALINES;
04   A A 5
05   A B 1
06   A C 0
07   B A 0
08   B B 3
09   B C 2
10   C A 1
11   C B 1
12   C C 2
13   ;
14   PROC FREQ DATA = RatingFile;
15        WEIGHT Count;
16        TABLES Rater1*Rater2/AGREE(WT=FC);
17        RUN;
```

Program 4.1. Basic SAS Program for Computing Kappa and FC-Based Weighted Kappa.

The Cicchetti-Allison (CA) Weights

The values taken by the CA weights depend on the ratings' data type, which could be of numeric or character types. It is essential for you to have a good understanding of how SAS creates the weights, in order to ensure that weighting is carried out properly. CA weights values are determined as follows:

▶ When the categories that appear in the SAS data step are of character type, then *SAS will first sort the categories* before numbering them sequentially from 1 to the number of categories. In Program 4.1 for example, the sorted categories A, B, and C will respectively be assigned the numbers 1, 2, and 3. The CA weight associated with the (A,B) pair of categories is determined as follows:

$$\text{Weight}[1,2] = 1 - \frac{|1-2|}{3-1} = 1 - \frac{|-1|}{2} = 1 - \frac{1}{2} = 1/2,$$

where $|-1|$ represents the absolute value of -1, which is 1. Note that the denominator $3-1$ in the above expression applies to all pairs of categories and represents the difference between the maximum and the minimum values.

The CA weights for the dataset of Program 4.1 are the following:

Table 4.1: CA Weights for Program 4.1 Data

	A	B	C
A	1.0	0.5	0.0
B	0.5	1.0	0.5
C	0.0	0.5	0.0

▶ When the categories that appear in the SAS data step are of numeric type, then SAS will use these numbers. Suppose that the raters can assign 3 possible numbers 1.8, 4.3, and 5.6 to subjects. The CA weight associated with the pair of scores 4.3 and 5.6 for example is calculated as follows:

$$\text{CA}[4.3, 5.6] = 1 - \frac{|4.3 - 5.6|}{5.6 - 1.8} = 1 - \frac{|-1.3|}{3.8} = 0.6579.$$

The whole set of CA weights associated with the set of scores $\{1.8, 4.3, 5.6\}$ is shown in Table 4.2.

Table 4.2: CA Weights for Scores 1.8, 4.3, and 5.6

	1.8	4.3	5.6
1.8	1.0	0.3421	0.0
4.3	0.3421	1.0	0.6579
5.6	0.0	0.6579	0.0

The Fleiss-Cohen (FC) Weights

The process for calculating FC weights is similar to that used to calculate CA weights.

4.2 The Weights.

▶ When categories are of character type *SAS will sort* and number them sequentially from 1 to the number of categories. In Program 4.1, the sorted categories A, B, and C will respectively be assigned the numbers 1, 2, and 3. The FC weight associated with the (A,B) pair of categories for example is determined as follows :

$$\text{Weight}[1,2] = 1 - \frac{(1-2)^2}{(3-1)^2} = 1 - \frac{1}{4} = 0.75,$$

The denominator $(3-1)^2 = 4$ in the above expression applies to all pairs of categories and represents the squared difference between the maximum and the minimum values.

FC weights for the dataset of Program 4.1 are the following :

Table 4.3: FC Weights for Program 4.1 Data

	A	B	C
A	1.0	0.75	0.0
B	0.75	1.0	0.75
C	0.0	0.75	0.0

▶ If categories are of numeric type, then SAS will use these numbers. Suppose that the raters can assign 3 possible numbers 1.8, 4.3, and 5.6 to the raters. The FA weight associated with the pair of scores 4.3 and 5.6 is calculated as follows:

$$\text{CA}[4.3, 5.6] = 1 - \frac{(4.3-5.6)^2}{(5.6-1.8)^2} = 1 - \frac{(-1.3)^2}{(3.8)^2} = 0.8830.$$

The whole set of FC weights associated with the scores {1.8, 4.3, 5.6} is shown in Table 4.4.

Table 4.4: CA Weights for Scores 1.8, 4.3, and 5.6

	1.8	4.3	5.6
1.8	1.0	0.5672	0.0
4.3	0.5672	1.0	0.8830
5.6	0.0	0.8830	0.0

Meaning of the Weights

CA and FC weights do not necessarily have any practical meaning. You will choose one set of weights based on the extent to which you want to downweight serious disagreements. The CA weight associated with the A-B disagreement in Program 4.1 is 0.5, while the FC weight associated with the same disagreement is 0.75. The choice of one weight over another can be largely subjective, and may be decided by a panel of experts. But using simple unweighted Kappa when there is a clear hierarchy among categories is likely to underestimate the extent of agreement.

WARNING

In Table 4.1 the pair (A,B) is assigned a CA weight of 0.5 while (A,C) is assigned a 0 weight. This indicates that if the 2 raters classified a subject into categories A and C, they will have shown a disagreement deemed far more serious (hence a low weight of 0) than if they classify into categories A, and B associated with a weight of 0.5. How does SAS determine which pair should be assigned a CA-weight of 0.5, and which one should get a 0 weight?

Consider as an example the categories "No Pain", "Mild Pain", "Moderate Pain", "Severe Pain". I do not need to be an expert in human pain to understand that a Moderate Pain is more acute than a No Pain. However SAS may not understand that basic fact as easily. Consider Program 4.2, and the results in Output 4.1. The weighted Kappa of 0.6863, which is based on CA weights is **wrong**. To understand why, you need to look at the ordering of categories in the contingency table (Mild, Moderate, No, and Severe). SAS calculates the weighted Kappa as if these categories were ordered by the intensity of the pain. Since the "No Pain" and "Severe Pain" categories are side by side, the No-Severe disagree-

4.2 The Weights.

ment is not considered very serious, and is assigned a **CA**-weight of $1 - |3 - 4|/(4 - 1) = 0.6667$, when in fact it is the most serious disagreement that should receive a 0 **CA**-weight.

In order to resolve this problem and compute the correct weighted Kappa, I recommend to replace Program 4.2 with Program 4.3. The main difference between these 2 programs is the use of the **FORMAT** procedure, where each category is assigned a numeric integer value that increases with the pain intensity.

The input dataset (RatingFile) is now built on these numeric values. That is, the **FREQ** procedure will now use these numeric values to define the **CA** and **FC** weights. These numbers need not be 1,2 3, and 4. They could be anything you want them to be.

> *When using the **FREQ** procedure to compute simple and weighted kappa coefficients, it is safer to use the **FORMAT** procedure to first assign to categories, numeric values that determine their rankings.*

If for example, the researcher believes that the intensity gap between a severe and a moderate pain is bigger than the gap between the moderate and the mild pain, then 4 could well be replaced by 4.5 or by 5 or anything larger.

Program 4.3 yields the results shown in Output 4.2. We can now see the new and correct ordering of the categories in the contingency table. But more important, the new and correct weighted kappa is now 0.8012, which is much higher than the one produced by program 4.2.

```
01  DATA RatingFile;
02      LENGTH Rater1$ 8 Rater2$ 8;
03      INPUT Rater1$ Rater2$ Count;
04      DATALINES;
05  No Mild 8
06  No No 20
07  Moderate Moderate 25
08  Moderate Mild 5
09  Severe Severe 30
10  Severe Moderate 10
11  Mild Mild 15
12  Mild Moderate 6
13  ;
14  PROC FREQ DATA=RatingFile;
15      WEIGHT Count;
16      TABLES Rater1*Rater2/AGREE NOPERCENT NOROW NOCOL;
17      RUN;
```

Program 4.2. Basic SAS Program for Computing Kappa based on a Contingency Table.

4.2 The Weights.

```
01   PROC FORMAT;
02      VALUE CategFmt
03         1 = 'No'
04         2 = 'Mild'
05         3 = 'Moderate'
06         4 = 'Severe';
07   RUN;
08   DATA RatingFile;
09      INPUT Rater1 Rater2 Count;
10      DATALINES;
11   1 2 8
12   1 1 20
13   3 3 25
14   3 2 5
15   4 4 30
16   4 3 10
17   2 2 15
18   2 3 6
19   ;
20   PROC FREQ DATA=RatingFile;
21      WEIGHT Count;
22      TABLES Rater1*Rater2/AGREE NOPERCENT NOROW NOCOL;
23      FORMAT Rater1 Rater2 CategFmt.;
24   RUN;
```

Program 4.3. Basic SAS Program for Computing Kappa based on a Contingency Table.

Output 4.1. Output of SAS Programs 1 and 2

The FREQ Procedure

Table 1 of Rater1 by Rater2

Rater1   Rater2

Frequency	Mild	Moderate	No	Severe	Total
Mild	15	6	0	0	21
Moderate	5	25	0	0	30
No	8	0	20	0	30
Severe	0	10	0	0	30
Total	28	41	20	30	30

Statistics for Table 1 of Rater1 by Rater2

Test of Symmetry

Statistic (S)   18.0909
DF   6
Pr > S   0.0060

Kappa Statistics

Statistic	Value	ASE	95% Confidence Limits	
Simple Kappa	0.6739	0.0517	0.5725	0.7753
Weighted Kappa	0.6863	0.0525	0.5834	0.7891

Sample Size = 119

4.2 The Weights.

Output 4.2. Output of SAS Programs 1 and 2

The FREQ Procedure

Table 1 of Rater1 by Rater2

Rater1   Rater2

Frequency	No	Mild	Moderate	Severe	Total
No	20	8	0	0	28
Mild	0	15	6	0	21
Moderate	0	5	25	0	30
Severe	0	0	10	30	40
Total	20	28	41	30	119

Statistics for Table 1 of Rater1 by Rater2

Test of Symmetry

Statistic (S)	18.0909
DF	6
Pr > S	0.0060

Kappa Statistics

Statistic	Value	ASE	95% Confidence Limits	
Simple Kappa	0.6739	0.0517	0.5725	0.7753
Weighted Kappa	0.8012	0.0340	0.7346	0.8679

Sample Size = 119

4.3 The AgreeStat_2SAS Macro

I wrote the AgreeStat_2SAS Macro[1] to accomplish the following two tasks :

▶ Allow the user to supply an input dataset of ratings without having to worry about the unbalanced-table and diagonal problems. This macro will take any input dataset of ratings and will always compute the correct simple and weighted Kappa.

▶ Allow the user to experiment with agreement coefficients other than Kappa. The alternative agreement coefficients considered in this version of the macro are Scott's Pi statistic, Brennan-Prediger, and Gwet's AC_1. The unweighted (simple) as well as the weighted versions of these alternative coefficients are offered along with associated standard errors, confidence intervals and P-values.

An Example

The AgreeStat_2SAS macro is a SAS program that is obtained in a separate file called AgreeStat_2SAS.sas. For this example, I am assuming that you have stored it in a directory called
c:\advancedanalytics\myproject
The goal here is to evaluate the extent of agreement between 2 judges named Judge 1 and Judge 2, using the simple and weighted versions of Cohen's Kappa, Scott's Pi, Brennan-Prediger's coefficient, and Gwet's AC_1. These calculations will be done using the following rating data:

---

[1]This macro can be downloaded at: www.agreestat.com/agreestat.html

4.3 The AgreeStat_2SAS Macro.   - 57 -

Table 4.5: Rating Data from Judges 1 and 2

Judge 1	Judge 2	Count
A	A	15
A	B	1
B	A	2
B	B	10
C	A	3
C	B	4

Program 4.4 shows how you may call the SAS macro AgreeStat_2SAS to analyze your rating data. This macro has 6 parameters that must be specified, some of which are mandatory while others are optional. I now like to review each of these parameters.

```
01   %LET pdir=c:\advancedanalytics\myproject;
02   DATA SkinCondition;
03     INPUT Judge1$ Judge2$ Count;
04     DATALINES;
05   A A 15
06   A B 1
07   B A 2
08   B B 10
09   C A 3
10   C B 4
11   ;
12   %INCLUDE ''&pdir\AgreeStat_2SAS.sas";
13   %AgreeStat_2SAS(
14        InputFile=SkinCondition,
15        Rater1=Judge1,
16        Rater2=Judge2,
17        CountVar=Count,
18        WeightType=FC,
19        CONFLEV=0.95);
```

Program 4.4. Illustration of the AgreeStat_2SAS Macro

The AgreeStat_2SAS's Parameters

- **InputFile** = This parameter is mandatory, and represents the name of the file containing the ratings. It must have a minimum of 2 variables associated with the rater's ratings. The file may contain other variables as well.

- **Rater1** = This parameter is mandatory, and is the variable name associated with the first rater's ratings.

- **Rater2** = This parameter is mandatory, and is the variable name associated with the second rater's ratings.

- **CountVar** = This parameter is optional. When present, it will contain the name of the numeric variable showing the frequency of occurrence of each category combination used by both raters. In the previous examples discussed in this book, it was the name "Count" that I used. When this parameter is not specified, the macro will create a variable Count, will add it to the input file, and assign a value of 1 to it. *Leave this parameter unspecified when using raw data, and specify it when using summary data.*

- **WeightType** = This parameter is optional. When left unspecified, the macro will use the CA-weights to compute the weighted agreement coefficients. Otherwise, it may take the following values :

    CA: for the Cicchetti-Allison weights (the default value)

    FC: for the Fleiss-Cohen weights

    RA: for Gwet's Radical Weights (if interested, see Gwet (2010) for a discussion of these weights)

    FILE_NAME: The macro considers any other name supplied here as a SAS dataset containing the user's custom weights. If 4 categories for example are used, then this file must have 4

4.3 The AgreeStat_2SAS Macro.

variables placed in alphabetical order in the file and 4 records also placed in alphabetical order (see Tables 4.3, and 4.4).

▶ **ConfLev** = This parameter is optional, and represents the confidence level used for deriving confidence intervals. When left unspecified, the macro uses the default value of 0.95. You may also specify any value between 0 and 1. The typical values for the confidence level are 0.90, 0.95, and 0.99.

The Macro's Output

Executing Program 4.4 will produce the results shown in Output 4.3. We have separated these results into 5 parts that are described as follows:

▶ **Part 1**
Part 1 of the macro's output is essentially the output from the FREQ procedure of SAS. The only difference being that with this macro the user does not have to worry about the diagonal and the unbalanced table issues. The subtitle "Controlling for FLAG=1" will appear only if the macro had to resolve either the unbalanced table or the diagonal problem.

▶ **Part 2**
Part 2 of the macro's output will appear only if the macro had to resolve the unbalanced-table or the diagonal issues. It is garbage that my macro generated after fixing the problem, but could not throw away for you. You can safely ignored it.

▶ **Part 3**
Part 3 of the macro's output shows a number of statistics for the unweighted versions of Kappa, Scott's Pi, Gwet's AC_1, and Brennan-Prediger (BP). These statistics are the following :
→ Coefficient: This is the inter-rater reliability coefficient.
→ Standard Error: This is the standard error of the coefficient.

→ **Lower Conf. Limit:** This is the lower bound of the confidence interval, calculated at the confidence level specified by you.
→ **Upper Conf. Limit:** This is the upper bound of the confidence interval, calculated at the confidence level specified by you.
→ **One-Sided P-Value:** The one-sided P-value is useful only if the user wants to test the hypothesis that the "true" coefficient is 0 or greater than 0. As a rule of thumb, I suggest to consider a one-sided p-value[2] that is smaller than 0.05 as a statistical evidence that the "true" coefficient is indeed bigger than 0. That is, the observed ratings show the existence of an intrinsic agreement among raters.
→ **Two-Sided P-Value:** The two-sided P-value is useful only if you want to test the hypothesis that the "true" coefficient is different from 0. As a rule of thumb, I suggest to consider a two-sided p-value that is smaller than 0.05 as a statistical evidence that the "true" coefficient is indeed different from 0. *Although the two-sided P-value is popular in the general statistical literature, it is less useful than the one-sided version in the study of inter-rater reliability.*
→ **Z-Value:** The Z-value is simply a ratio of a coefficient to its standard error. It is used for computing the P-value, and does not present any particular interest, except to users who may want to do further analyses with it.

▶ **Part 4**

Part 4 of the macro's output is very similar to Part 3, except that it is about the weighted coefficients. The standard error associated with the weighted kappa shown here will generally be slightly different from that produced by the **FREQ** procedure of **SAS**. This is due to the fact **SAS** has implemented

---

[2] A P-value = 0.0000 simply means that its value is smaller than 0.0001

4.3 The AgreeStat_2SAS Macro.

the standard error formulas proposed by Fleiss et al. (1969), while I have implemented the standard variance calculation based on the jackknife methodology. When the number of observation is limited, which is typically the case in the overwhelmingly majority of reliability studies, there is no need to used complicated equations to obtain standard errors when the jackknife methodology offers an easily implementable approach (see Gwet (2010) for further discussion on this topic).

▶ **Part 5**

Part 5 of the macro's output shows the specific weights used to compute the weighted coefficients.

Output 4.3. Output of SAS Program 4.4
(Part 1)

```
              AgreeStat-2SAS
Simple & Weighted Agreement Coefficients (Weights: FC)

              The FREQ Procedure

         Table 1 of Judge1 by Judge2
           Controlling for FLAG=1
         Judge1     Judge2
```

Frequency	A	B	C	Total
A	15	1	0	16
B	2	10	0	12
C	3	4	0	7
Total	20	15	0	35

```
    Statistics for Table 1 of Judge1 by Judge2
              Controlling for FLAG=1

              Test of Symmetry
           Statistic (S)   7.3333
           DF                   3
           Pr > S          0.0620

              Kappa Statistics
```

Statistic	Value	ASE	95% Confidence Limits	
Simple Kappa	0.5172	0.1076	0.3063	0.7281
Weighted Kappa	0.4192	0.1380	0.1488	0.6896

Sample Size = 35

4.3 The AgreeStat_2SAS Macro.   - 63 -

(Part 2)

```
               AgreeStat-2SAS
Simple & Weighted Agreement Coefficients (Weights: FC)

                The FREQ Procedure

            Table 1 of Judge1 by Judge2
                Controlling for FLAG

              Overall Kappa Statistics
Statistic         Value    ASE    95% Confidence Limits
Simple Kappa      0.5172  0.1076   0.3063      0.7281
Weighted Kappa    0.4192  0.1380   0.1488      0.6896

         Tests for Equal Kappa Coefficients
       Statistic       Chi-Square  DF   Pr > ChiSq
       Simple Kappa      0.0000     0      .
       Weighted Kappa    0.0000     0      .

            Effective Sample Size = 35
              Frequency Missing = 9

      WARNING: 20% of the data are missing
```

(Part 3)

```
                  AgreeStat-2SAS
        Simple (or Unweighted) Agreement Coefficients
Statistic                Kappa    Scott    AC₁     BP
Coefficient              0.5172   0.5046   0.5985  0.5714
Standard Error           0.1076   0.1160   0.1135  0.1145
95% Lower Conf. Limit    0.3063   0.2772   0.3761  0.3469
95% Upper Conf. Limit    0.7281   0.7320   0.8209  0.7959
One-Sided P-Value        0.0000   0.0000   0.0000  0.0000
Two-Sided P-Value        0.0000   0.0000   0.0000  0.0000
Z-Value                  4.8068   4.3484   5.2745  4.9889
```

Chapter 4. *Weighted Kappa with Proc FREQ*

(Part 4)

AgreeStat-2SAS
Weighted Agreement Coefficients (Weights: FC

Statistic	Kappa	Scott	AC_1	BP
Coefficient	0.4192	0.3868	0.6514	0.5929
Standard Error	0.1489	0.1697	0.1337	0.1453
95% Lower Conf. Limit	0.1273	0.0542	0.3894	0.3080
95% Upper Conf. Limit	0.7111	0.7194	0.9135	0.8777
One-Sided P-Value	0.0024	0.0113	0.0000	0.0000
Two-Sided P-Value	0.0024	0.0113	0.0000	0.0000
Z-Value	2.8150	2.2796	4.8719	4.0792

(Part 5)

AgreeStat−2SAS
The Weights: FC

1.00	0.75	0.00
0.75	1.00	0.75
0.00	0.75	1.00

Some Remarks on the Macro

▶ All agreement coefficients implemented in this SAS macro can be used with ordinal and interval scores (or categories). The categories are ordinal if the scores the raters may assign to subjects can be ordered (or ranked) somehow from low to high levels. Interval scores are ordinal as well, and allow for basic arithmetic operations such as addition and subtraction. For such scores, I strongly recommend to supply numeric values to SAS in the data step. This is the most effective way to get SAS produce correct weighted coefficients, due to the natural way numeric values can be ordered.

4.3 The AgreeStat_2SAS Macro.

- ▶ If you decide to use ordinal categories but do not want to use numeric values, then my suggestion to you is to left-pad your category names with a 2-digit number such as 01, 02, 03, and so on, according to the hierarchical order of these categories. For example you will have `01NoPain`, and `02MildPain` to indicate that a mild pain is more acute than a no pain.
- ▶ All coefficients discussed by Gwet (2010) for 2 raters and ordinal/interval data in chapter 2, and in parts of chapter 4 (including the AC_2) can be calculated using this macro. But one needs to specify the FC weights, and numeric values in the data step.
- ▶ The standard errors associated with the weighted coefficients are usually based on complex equations. For the weighted Kappa coefficient, SAS uses the equations proposed by Fleiss et al. (1969) to obtain the standard errors. These equations work fine. In my macro, I have used the jackknife approach[3]. The jackknife standard error is generally close to that based on Fleiss' equation, and has the advantage that it can easily be applied to other agreement coefficients.

If the inter-rater reliability is based on 10 subjects for example, the standard error by jackknife is calculated by first calculating a series of 10 agreement coefficients each obtained after removing one subject. It is the observed variation in these replicated coefficients that will lead to the jackknife standard error.

---

[3]This approach is very popular in the statistical community, and provides a general and standard way for computing standard errors of almost any statistic (see Gwet (2010) for further discussion)

4.4 Testing Kappa for Statistical Significance

In addition to computing the simple and weighted Kappa statistics, you may also want to known whether the obtained estimates are statistically significant. The question is whether simple and weighted Kappas are statistically different from 0. The problem amounts to testing whether the extent of agreement among raters is not merely the result of sampling variability, which would mean that our coefficient does not represent "intrinsic" agreement. SAS offers a solution to this problem[4] with the **TEST** statement of the **FREQ** procedure as shown in Program 4.5.

```
01  DATA RatingFile;
02     INPUT Rater1 Rater2;
03     DATALINES;
04  1 2
05  1 1
06  3 3
07  2 2
08  1 1
09  2 2
10  1 1
11  2 2
12  1 3
13  1 3
14  ;
15  PROC FREQ DATA = RatingFile;
16     TABLES Rater1*Rater2/AGREE;
17     TEST AGREE;
18     RUN;
```

Program 4.5. Testing Kappa for Statistical Significance

Running this program produces the results shown in Output 4.4. In

---

[4]Researchers typically want Kappa to be the highest possible. Therefore, I personally think that a mere statistical test of significance may not be a so powerful tool after all. It only compares Kappa to 0, and may only prove that your study is not terribly bad.

4.4 Testing Kappa for Statistical Significance.

addition to the standard results of the **FREQ** procedure discussed previously, you will get the one-sided and two-sided p-values that we also discussed in previous sections. Note that if the **AgreeStat_2SAS** macro is used, these P-values will automatically be calculated.

If the user wants to test the statistical significance of the simple Kappa only, and not that of the weighted Kappa, then it can be achieved by replacing line 17 in Program 4.5 with **TEST KAPPA**;.

In order to conduct a significance test on the weighted Kappa only and not on the simple kappa, one should replace line 17 in Program 4.5 with **TEST WTKAP**; *These are minor options that* **SAS** *offers.*

Output 4.4. Output of SAS Program 4.5
(Part 1)

```
                    The FREQ Procedure

               Table of Rater1 by Rater2
    Rater1        Rater2

    Frequency
    Percent
    Row Pct
    Col Pct         1  |     2  |     3  | Total
    ----------------+--------+--------+--------+
    1               3  |     1  |     2  |     6
                40.00 |  10.00 |  10.00 |  60.00
                66.67 |  16.67 |  16.67 |
               100.00 |  25.00 |  50.00 |
    ----------------+--------+--------+--------+
    2               0  |     3  |     0  |     3
                 0.00 |  30.00 |   0.00 |  30.00
                 0.00 | 100.00 |   0.00 |
                 0.00 |  75.00 |   0.00 |
    ----------------+--------+--------+--------+
    3               0  |     0  |     1  |     1
                 0.00 |   0.00 |  10.00 |  10.00
                 0.00 |   0.00 | 100.00 |
                 0.00 |   0.00 |  50.00 |
    ----------------+--------+--------+--------+
    Total           3  |     4  |     3  |    10
                40.00 |  40.00 |  20.00 | 100.00
```

Statistics for Table of Rater1 by Rater2

Test of Symmetry	
Statistic (S)	3.0000
DF	3
Pr > S	0.3916

4.4 Testing Kappa for Statistical Significance.

(Part 2)

Simple Kappa Coefficient	
Kappa	0.5522
ASE	0.1922
95% Lower Conf Limit	0.1756
95% Upper Conf Limit	0.9289

Test of H0: Kappa = 0

ASE under H0	0.2007
Z	2.7509
One-sided Pr > Z	0.0030
Two-sided Pr > |Z|	0.0059

Statistics for Table of Rater1 by Rater2

Weighted Kappa Coefficient	
Weighted Kappa	0.4318
ASE	0.2275
95% Lower Conf Limit	-0.0140
95% Upper Conf Limit	0.8776

Test of H0: Weighted Kappa = 0

ASE under H0	0.2034
Z	2.1229
One-sided Pr > Z	0.0169
Two-sided Pr > |Z|	0.0338

Sample Size = 10

Exact P-values

[5]The P-values and Z scores associated with simple and weighted kappa statistics are based on Asymptotic Standard Errors (ASE), which will be invalid when the number of subjects is too small. Fortunately, the **FREQ** procedure of **SAS** provides the option for computing exact P-values with the "**EXACT**" statement. The exact P-value is essentially a P-value that is calculated without assuming that the number of subjects is large. It is expected to yield more precise numbers when the number of subjects is very small.

The **EXACT** statement must be followed by one of the following three keywords:

▶ AGREE

This produces exact P-values for simple and weighted kappa coefficients

▶ KAPPA

This produces exact P-values for simple Kappa coefficient only.

▶ WTKAP

This produces exact P-values for weighted kappa coefficients only.

Program 4.6 will produce the results where the sections entitled "Simple Kappa Coefficient" and "Weighted Kappa Coefficient" would appear as shown in Output 4.5. These results show that exact P-values may occasionally be quite different from their asymptotic

---

[5]This section may be useful only to practitioners who have a handful of subjects (fewer than 10) and want to test Kappa for statistical significance. I personally advise against using only a few subjects to prove the reliability of a data collection procedure. The validity obtained with exact tests represents only statistical validity. Because the body of information our opinion is based upon remains slim, statistical validity cannot demonstrate the validity of the entire scientific procedure

4.4 Testing Kappa for Statistical Significance. - 71 -

approximations. The computation of exact P-values does not take too long when the subject sample of small or moderate size. Therefore, I recommend to always specify the EXACT statement, if you do want to analyze reliability data based on 15 subjects or fewer.

Even if the number of subjects is large and the user is still concerned about the accuracy of the asymptotic standard errors, the **FREQ** procedure provides the Monte Carlo option to the **EXACT** statement. The Monte-Carlo option instructs **SAS** not to use the network algorithm of Mehta and Patel (1983), but rather to generate several random contingency tables with the same marginal totals as the observed table.

```
01  DATA RatingFile;
02     INPUT Rater1 Rater2;
03     DATALINES;
04  1 2
05  1 1
06  3 3
07  2 2
08  1 1
09  2 2
10  1 1
11  2 2
12  1 3
13  1 3
14  ;
15  PROC FREQ DATA = RatingFile;
16     TABLES Rater1*Rater2/AGREE;
17     TEST AGREE;
18     EXACT AGREE;
19     RUN;
```

Program 4.6. Using Exact P-values

The Monte-Carlo method is used when the **MC** option is specified with the **EXACT** statement. By default **SAS** generates 10,000 random tables. The option "N=" allows you to change that number. "N=100,000" for example will generate 100,000 tables.

Output 4.5. Output of SAS Program 4.6

Simple Kappa Coefficient	
Kappa (K)	0.5522
ASE	0.1922
95% Lower Conf Limit	0.1756
95% Upper Conf Limit	0.9289

Test of H0 : Kappa = 0

ASE under H0	0.2007
Z	2.7509
One-sided Pr > Z	0.0030
Two-sided Pr > |Z|	0.0059

Exact Test
One-sided Pr >= K	0.0143
Two-sided Pr >= |K|	0.0143

Weighted Kappa Coefficient	
Weighted Kappa	0.4318
ASE	0.2275
95% Lower Conf Limit	-0.0140
95% Upper Conf Limit	0.8776

Test of H0 : Weighted Kappa = 0

ASE under H0	0.2034
Z	2.1229
One-sided Pr > Z	0.0169
Two-sided Pr > |Z|	0.0338

Exact Test
One-sided Pr >= K	0.0714
Two-sided Pr >= |K|	0.1012

Sample Size = 10

4.4 Testing Kappa for Statistical Significance.

Program 4.7 shows how the Monte-Carlo option can be used with the EXACT statement. In this example, I have added another option ALPHA=0.10. Its role is to specify the confidence level used to construct confidence interval around the Monte-Carlo P-value. Output 4.6 - Part 1 shows only the "Simple Kappa Coefficient" section of the output generated by Program 4.7, while Output 4.6 - Part 2 shows the "Weighted Kappa Coefficient" section.

In both parts 1 and 2, the statistics under the title "Monte-Carlo Estimates for the Exact Test" are the novelty. The first three statistics show the Monte-Carlo one-sided P-value under the label "Estimate", the 95% lower and upper confidence limits for the P-value. The closer the two confidence limits, the more precise the Monte-Carlo P-value. If all Monte-Carlo P-values are positive, then the one-sided and two-sided P-values will be identical. That is the situation we are in with simple Kappa (see Part 1 of the Output). The last number is the "Initial Seed." This number is useful only for obtaining the same results each time the program is run. The SEED option should then be used with the EXACT statement for that purpose. A statement such as the following :

> EXACT AGREE/MC N=10000 ALPHA=0.05
> SEED=49545 ;

will always yield the same results of Output 4.6.

```
01    PROC FREQ DATA = RatingFile;
02       TABLES R1*R2/AGREE;
03       TEST AGREE;
04       EXACT AGREE/MC N=10000 ALPHA=0.05;
05    RUN;
```

Program 4.7. Computing Exact Monte-Carlo P-values

Output 4.6. Output of SAS Program 4.6 (Part 1)

Simple Kappa Coefficient	
Kappa (K)	0.5522
ASE	0.1922
95% Lower Conf Limit	0.1756
95% Upper Conf Limit	0.9289

Test of H0: Kappa = 0

ASE under H0	0.2007
Z	2.7509
One-sided Pr > Z	0.0030
Two-sided Pr > |Z|	0.0059

Monte Carlo Estimates for the Exact Test
One-sided Pr >= K

Estimate	0.0140
95% Lower Conf Limit	0.0117
95% Upper Conf Limit	0.0163

Two-sided Pr >= \|K\|

Estimate	0.0140
95% Lower Conf Limit	0.0117
95% Upper Conf Limit	0.0163

Number of Samples	10000
Initial Seed	49545

4.4 Testing Kappa for Statistical Significance.

Output of SAS Program 4.6 (Part 2)

```
              The FREQ Procedure
         Weighted Kappa Coefficient
     Weighted Kappa (K)         0.4318
     ASE                        0.2275
     95% Lower Conf Limit      -0.0140
     95% Upper Conf Limit       0.8776


     Test of H0: Weighted Kappa = 0

     ASE under H0               0.2034
     Z                          2.1229
     One-sided Pr > Z           0.0169
     Two-sided Pr > |Z|         0.0338

Monte Carlo Estimates for the Exact Test
     One-sided Pr >= K
     Estimate                   0.0724
     95% Lower Conf Limit       0.0673
     95% Upper Conf Limit       0.0775


     Two-sided Pr >= |K|
     Estimate                   0.1028
     95% Lower Conf Limit       0.0968
     95% Upper Conf Limit       0.1088


     Number of Samples           10000
     Initial Seed            608612614
          Sample Size = 10
```

CHAPTER 5

Kappa for Multiple Raters with SAS

5.1 Introduction

As mentioned earlier, the **FREQ** procedure of SAS® can only compute the extent of agreement among two raters. When the number of raters is three or more, no **SAS** procedure has been developed to compute the Kappa's extensions that were proposed in the literature. However, one of these generalizations due to Fleiss (1971) can be calculated using a **SAS** macro called "magree.sas", and which is offered by **SAS** institute as a courtesy service to interested **SAS** users. This SAS macro can be downloaded at:

**http://support.sas.com/kb/25/addl
/fusion25006_1_magree.sas.txt**

This macro is well documented[1], and implements Fleiss' generalized Kappa as the only agreement coefficient applicable to 3 raters or more. In this chapter, I will focus on another **SAS** macro called **AgreeStat_3SAS**.sas that I have developed and that can compute Fleiss' generalized Kappa as well as a few other agreement statistics proposed in the literature.

---
[1]Interested user may find more here: http://support.sas.com/kb/25/006.html

More on the magree.sas SAS Macro

In addition to computing Kappa for multiple raters, the magree.sas macro can compute the Kendall's coefficient of concordance (see Siegel & Castellan Jr. (1988) for more on this coefficient) if the response variable is numeric. This macro also has the special feature of computing Kappa statistics conditionally on the response category. That is for each category, the conditional Kappa is computed as a measure of the extent of agreement between raters with respect to that specific category. The methods for computing these conditional kappa coefficients are also discussed by Fleiss (1971).

If you care about precision measures, then you may want to know that the version of the MAGREE.SAS macro (version 1.0) that I obtained does not in my opinion compute the variance of the conditional Kappa properly as specified by Fleiss (1971). In fact the correct variance formula is given by equation [23] in Fleiss' paper. Its implementation in MAGREE.SAS is erroneous although the standard errors obtained are often close to the correct estimations. However, the variance of the overall Kappa was properly implemented according to Fleiss (1971).

5.2 Agreement Among 3 Raters or More: A Review

I consider in this chapter a simple inter-rater reliability experiment where 4 raters named Rater1, Rater2, Rater3 and Rater4 must each classify 10 subjects into one of 5 possible categories named 1, 2, 3, 4, and 5. You have 2 options for organizing the data. The first option is to record the raw data as shown in Table 5.1. This format shows each rater and the specific category where it classified each subject. Its main advantage is to capture all the information related to the experiment. The second option is to display the distribution of raters by subject and category as shown in Table 5.2. This format has the advantage of giving you a quick

5.2. Agreement Among 3 Raters or More.

view of the way the raters scored the subjects. On the other hand, you cannot know how a rater categorized a given subject unless all 4 agreed by classifying the subject into the exact same category.

Table 5.1: Classification of 10 Subjects by 4 Raters in 5 Categories

Subject	Rater 1	Rater 2	Rater 3	Rater 4
1	5	5	5	5
2	3	1	3	1
3	5	5	5	5
4	3	1	3	1
5	5	5	4	5
6	1	2	3	3
7	3	1	1	1
8	1	1	1	3
9	3	3	4	4
10	1	3	1	1

Table 5.2: Distribution of Raters by Subject and Category

Subject	1	2	3	4	5	Total
1	0	0	0	0	4	4
2	2	0	2	0	0	4
3	0	0	0	0	4	4
4	2	0	2	0	0	4
5	0	0	0	1	3	4
6	1	1	2	0	0	4
7	3	0	1	0	0	4
8	3	0	1	0	0	4
9	0	0	2	2	0	4
10	3	0	1	0	0	4

My advice is to always have the raw data (Table 5.1) at our disposal whenever possible. Actually some interesting agreement coefficients proposed in the literature cannot be calculated at all if only Table 5.2 was available. After all you can always produce Table 5.2 using Table 5.1. The other way around would be impossible.

All agreement coefficients presented in this chapter share the same Kappa-like form, which presents itself as follows:

$$\text{COEFFICIENT} = \frac{\text{AP} - \text{CAP}}{1 - \text{CAP}},$$

where AP and CAP stand for Agreement Probability, and Chance-Agreement Probability respectively. Moreover, in addition to having the same general form, these coefficients will also use the same AP, and will only differ in the way the CAP is calculated. I will now show you step by step using Table 5.2, how the common AP is calculated:

How to Compute AP ?

Step 1. Cell-Level Agreement

If c is the content of a Table 5.2 cell then the likelihood that 2 arbitrary raters will agree on that cell (defined by the corresponding subject and category) is calculated as follows :

$$\left(\frac{c \times (c - 1)}{4 \times (4 - 1)} \right),$$

where 4 represents the number of raters.

Step 2. Subject-Level Agreement

All cell-level agreement propensities of step 1 must be summed separately for each row of Table 5.2 (i.e. each subject) to obtain 10 subject-level agreement propensities.

5.2. Agreement Among 3 Raters or More.

> **Step 3. The AP**
>
> The average of the 10 subject-level agreement propensities of step 2 is the Agreement probability AP needed to compute KAPPA-F.

Table 5.3: Calculation of the Agreement Probability (AP)

Subject	1	2	3	4	5	Total
1	0	0	0	0	1	1
2	0.1667	0	0.1667	0	0	0.3333
3	0	0	0	0	1	1
4	0.1667	0	0.1667	0	0	0.3333
5	0	0	0	0	0.5	0.5
6	0	0	0.1667	0	0	0.1667
7	0.5	0	0	0	0	0.5
8	0.5	0	0	0	0	0.5
9	0	0	0.1667	0.1667	0	0.3333
10	0.5	0	0	0	0	0.5
Average						**0.5167**

Fleiss' Generalized Kappa

Fleiss (1971) extended Cohen's Kappa to experiments involving 3 raters or more. I will denote Fleiss' generalized Kappa as KAPPA-F[2]. It maintains the same original form of Cohen's Kappa. That is,

$$\text{KAPPA-F} = \frac{\text{AP} - \text{CAP-F}}{1 - \text{CAP-F}},$$

where CAP-F stands for Chance-Agreement Probability according to Fleiss. Table 5.2 contains all the information needed to compute KAPPA-F, and will be used to describe the computation steps.

---
[2] Here letter F refers to Fleiss' version of the generalized Kappa, we will see other versions as well.

How to Compute CAP·F

[1] Cell-Level Relative Number of Raters
Let c be the content of a Table 5.2 cell. For each cell, compute the ratio c/4 (the relative number of raters in cell c)

[2] Category-Level Classification Probabilities (CCP)
Average the relative number of raters of step 1, separately for each category. *(This yields category-level classification probabilities as shown in Table 5.4)*

[3] Category-Level Chance-Agreement Probabilities (CCAP)
Square each of the category-level classification probabilities obtained in step 2.

[4] CAP·F
Sum all category-level chance-agreement probabilities of step 3 across categories to obtain the CAP·F.

Table 5.4: Distribution of Raters by Subjects and Category

Subject	1	2	3	4	5	Sum
1	0	0	0	0	1	1
2	0.5	0	0.5	0	0	1
3	0	0	0	0	1	1
4	0.5	0	0.5	0	0	1
5	0	0	0	0.25	0.75	1
6	0.25	0.25	0.5	0	0	1
7	0.75	0	0.25	0	0	1
8	0.75	0	0.25	0	0	1
9	0	0	0.5	0.5	0	1
10	0.75	0	0.25	0	0	1
CCP	0.35	0.025	0.275	0.075	0.275	1
CCAP	0.1225	0.000625	0.075625	0.005625	0.075625	**0.28**

5.2. Agreement Among 3 Raters or More.

Fleiss' generalized Kappa agreement coefficient can now be calculated as follows :

$$\boxed{\text{KAPPA·F} = \frac{0.5167 - 0.28}{1 - 0.28} = 0.3287.}$$

Conger's Generalized Kappa

Conger (1980) criticized Fleiss's coefficient on the ground that it does not reduce to Cohen's Kappa when the number of raters is 2. Conger's point is pertinent because the initial goal set by Fleiss was well to extend Kappa to the more general situation of 3 raters or more. Although Conger's generalized Kappa requires a little more computations, it represents an authentic extension of Cohen's 2-rater Kappa[3].

Conger has adopted the same general form of Kappa involving the AP and the CAP, with only the CAP differing from what Fleiss has proposed. Let us refer to Conger's specific CAP as CAP·C.

How to Compute CAP·C

[1] Compute CAP·F
Start by computing CAP·F as described previously.

[2] Distribute Subjects by Rater and Category
Using Table 5.1, create Table 5.5 showing the distribution of subjects by rater and category. The table cell associated with rater R and category C will show the number of subjects that rater R categorized into C.

---

[3]Conger's coefficient indeed reduces to Cohen's Kappa when the number of raters is limited to 2

Table 5.5: Distribution of Subjects by Rater and Category

Rater	1	2	3	4	5	Total
Rater1	3	0	4	0	3	10
Rater2	4	1	2	0	3	10
Rater3	3	0	3	2	2	10
Rater4	4	0	2	1	3	10

[3] Variation of the Raters' Classification Propensities (VCP)

- ▶ For each Table 5.5 cell c, compute the relative number as c/10, 10 being the number of subjects.
- ▶ For each column separately, compute the variance of the relative numbers so obtained (see Table 5.6)
- ▶ Sum all column-level variances to obtain **VCP** = 0.0267 as shown in Table 5.6 below.

Table 5.6: Calculation of the VCPs

Rater	1	2	3	4	5	Total
Rater1	0.3	0	0.4	0	0.3	1
Rater2	0.4	0.1	0.2	0	0.3	1
Rater3	0.3	0	0.3	0.2	0.2	1
Rater4	0.4	0	0.2	0.1	0.3	1
Variance	0.0033	0.0025	0.0092	0.0092	0.0025	0.0267

[4] Compute CAP·C

Finally, Conger's chance agreement probability is obtained as follows:

$$\text{CAP·C} = \text{CAP·F} - \text{VCP}/4 = 0.28 - 0.0267/4 = 0.2733,$$

where **CAP·F** is from step 1, and **VCP** from step 3.

Conger's generalized Kappa coefficient can now be calculated as

5.2. Agreement Among 3 Raters or More.

follows:

$$\text{KAPPA·C} = \frac{0.5167 - 0.2733}{1 - 0.2733} = 0.3349.$$

Brennan-Prediger Coefficient

The Brennan-Prediger coefficient, the simplest of all chance-corrected agreement coefficients is obtained as follows:

$$\begin{aligned}\text{BP} &= \frac{\text{AP} - 1/(\#\ \text{Categories})}{1 - 1/(\#\ \text{Categories})}, \\ &= \frac{0.5167 - 0.2}{1 - 0.2} \quad \text{(Using Table 5.2 data)}, \\ &= 0.3958.\end{aligned}$$

Gwet's AC_1 Coefficient

Gwet (2008a) introduced the AC_1 statistic as a paradox-robust alternative to Kappa, for evaluating the extent of agreement among raters. For a detailed discussion of AC_1, one could read Gwet (2008a), and Gwet (2010). This coefficient also has the general form of Kappa, which depends on the AP and the CAP. The AP is identical to that used for Conger and Fleiss generalized Kappa. However, the CAP is specific to AC_1, is denoted by CAP·G and is calculated as follows:

> [1] Compute CAP·F
> As a first step, you need to compute CAP·F, the CAP recommended by Fleiss (1971) as described earlier in this section.

[2] Compute CAP·G
Gwet's CAP is then obtained as follows:

$$\text{CAP·G} = (1 - \text{CAP·F})/(\#\text{Categories} - 1)$$

[3] Application
Using Table 5.2 data, CAP·G is calculated as follows:

$$\text{CAP·G} = (1 - 0.28)/(5 - 1) = 0.18.$$

Consequently, $AC_1 = (0.5167 - 0.18)/(1 - 0.18) = 0.4106.$

Gwet's AC_{1C} Coefficient

If the 2-rater form of AC_1 is extended to 3 raters or more using the same techniques Conger (1980) used to extend Cohen's Kappa to multiple raters, then one obtains a different form of AC_1 that is denoted by AC_{1C} (see Gwet (2010) for more details). AC_{1C} also uses the same AP as the other coefficients. Conger's approach for generalizing to 3 raters or more is to compute the chance-agreement probability (CAP) by averaging all pairwise chance-agreement probabilities[4]. Conger's approach ensures that the generalized expression always reduces to the initial 2-rater version for 2 raters.

CAP·GC is obtained as follows:

$$\text{CAP·GC} = \frac{(\text{NR}/2) \times (1 - \text{CAP·F}) - (\text{NR}/2 - 1) \times (1 - \text{CAP·C})}{\text{NC} - 1},$$

where NR = Number of Raters, and NC = Number of Categories.

---

[4]For each pair of raters, CAP is obtained with the same method described earlier in chapter 2 for two raters

5.2. Agreement Among 3 Raters or More.

Therefore, AC_{1C} is calculated as follows:

$$AC_{1C} = \frac{AP - CAP \cdot GC}{1 - CAP \cdot GC}$$

Using previous results one obtains the following:

$$CAP \cdot GC = \frac{(4/2) \times (1 - 0.28) - (4/2 - 1)(1 - 0.2733)}{5 - 1} = 0.1783$$

This leads to

$$AC_{1C} = (0.5167 - 0.1783)/(1 - 0.1783) = 0.4118.$$

Scott's Generalized Coefficient: Kappa·FC

This coefficient, denoted by **Kappa·FC** was obtained by generalizing the **PI** statistic of Scott (1955) for 2 raters to 3 raters or more using Conger's technique. The **AP** common to all agreement coefficients is still used, but the **CAP** is calculated by averaging all pairwise **CAP**'s (see chapter 2 for the method to calculate Scott's **CAP** for 2 raters). This new multiple-rater version of Scott's **CAP** is denoted **CAP·FC**[5]

CAP·FC is obtained as follows:

$$CAP \cdot FC = (NR/2) \times CAP \cdot F - (NR/2 - 1) \times CAP \cdot C$$

Therefore, **Kappa·FC** is calculated as follows:

$$Kappa \cdot FC = \frac{AP - CAP \cdot FC}{1 - CAP \cdot FC}$$

---

[5]The acronym FC stands for Fleiss and Conger, and indicates that the generalization techniques advocated by both authors are used to derive this chance-agreement probability

Using Table one obtains the following:

$$\text{CAP·FC} = (4/2) \times 0.28 - (4/2 - 1) \times 0.2733 = 0.2867$$

This leads to the following expression:

$$\text{AC}_{1\text{C}} = (0.5167 - 0.2867)/(1 - 0.2867) = 0.3224.$$

All 6 multiple-rater agreement coefficients presented in this section can be weighted using Cicchetti-Allison, Fleiss-Cohen, or the Radical weights. The weighted versions of all these coefficients are not presented here. However, interested readers could find a detailed presentation of these weighted coefficients in Gwet (2010). In the next section, I will present the SAS macro AgreeStat_3SAS that implements all 6 coefficients as well as their weighted versions as an option.

5.3 The AgreeStat_3SAS Macro

AgreeStat_3SAS is a SAS macro that I developed to compute the 6 agreement coefficients discussed in this chapter as well as their weighted versions, and associated standard errors and P-values. This macro can be downloaded at:
www.agreestat.com/agreestat.html
I developed it using Base SAS as well as the SAS/IML software.

Program 5.1 is a small SAS program that analyzes ratings collected during a reliability study of movement-related pain on 6 patients. Three raters classified the 6 patients into one of the 3 categories "No Pain," "Moderate Pain," and "Severe Pain." This program, which produces the results displayed in Output 5.1, will be described further in details later in this section. This output

5.3 The AgreeStat_3SAS Macro.

is rather lengthy because in addition to the 6 coefficients presented earlier, the program also performed the weighted analysis using Quadratic (or FC), Linear (or CA), and Radical weights. The user may decide not to request the weighted analysis if weighting is deemed of no interest. I will first described the different output components, before explaining how it was obtained so that the user can tailor the macro operation to fit specific needs.

```
01  %LET pdir=c:\advancedanalytics\myproject;
02  DATA CategoryFile;
03     LENGTH CategName$ 15;
04     INPUT CategName;
05     DATALINES;
06  No_Pain
07  Moderate_Pain
08  Severe_Pain
09  ;
10  DATA RatingFile;
11     LENGTH Rater1-Rater3 $15;
12     INPUT Rater1 Rater2 Rater3;
13     DATALINES;
14  No_Pain No_Pain No_Pain
15  Severe_Pain Severe_Pain Moderate_Pain
16  No_Pain Moderate_Pain No_Pain
17  Severe_Pain Severe_Pain Severe_Pain
18  Moderate_Pain Severe_Pain Severe_Pain
19  Moderate_Pain Moderate_Pain Moderate_Pain
20  ;
21  %INCLUDE "&pdir\AgreeStat_3SAS.sas";
22  %AgreeStat(
23       InputFile = RatingFile,
24       CategFile = CategoryFile,
25       InputType = r,
26       Weighted  = y);
```

Program 5.1. Illustration of the AgreeStat_2SAS Macro

The Macro's Output

The Distribution of Subjects by Rater and Category

The first output table shows the distribution of subjects by rater and category. One can see from this table that Rater2 classified 3 subjects into the "Severe_Pain" category. This table has no purpose other than providing a condensed summary of the ratings. It may help you perform a quick verification of the data that the macro has processed.

The Unweighted Analysis

The unweighted analysis provides estimates of the inter-rater reliability using the 6 coefficients presented earlier in this chapter. Are presented in this table, the coefficients, their standard errors, the P-values, the Unconditional Standard Error (U_StdErr)[6], and finally the Unconditional P-value, which is calculated based on the Unconditional Standard Error.

Note that Unconditional Standard Errors, and Unconditional P-values are missing for the coefficients **AC1·C**, **Kappa·C**, and **AC1·FC**. It is because no method for computing unconditional standard error for these coefficients is available at the present time.

Typically, a p-value that is smaller than 0.05 indicates that the corresponding agreement coefficient is "statistically significant." That is, the calculated coefficient has a magnitude that exceeds the maximum value that sampling error alone can produce. In this case, you may claim that your rating data has demonstrated the existence of an intrinsic agreement among raters beyond chance.

---

[6]This standard error is always bigger than the simple standard error as it combines the variation due to both the subjects and the raters (see Gwet (2010) for a more detailed discussion on this topic)

5.3 The AgreeStat_3SAS Macro.

Output 5.1. Output of SAS Program 5.1

```
            AgreeStat_3SAS (Ver 1.0)
     INTER−RATER RELIABILITY ANALYSIS RESULTS
   Distribution of Subjects by Rater and Category
                     Category
       Rater           MODERATE_   SEVERE_
       Names   NO_PAIN    PAIN      PAIN
       Rater1     2        2         2
       Rater2     1        2         3
       Rater3     2        2         2
```

```
            AgreeStat_3SAS (Ver 1.0)
     INTER−RATER RELIABILITY ANALYSIS RESULTS
   Unweighted Inter-Rater Reliability Coefficients,
      Associated Standard Errors, and P-values
                                   U_Std
Method    Estimate  StdErr  Pvalue   Err    U_Pvalue
AC1        0.50230  0.22093 0.011497 0.22097 0.011507
Kappa·F    0.49533  0.23157 0.016219 0.23170 0.016267
Br.−Pred.  0.50000  0.22361 0.012674 0.22361 0.012674
AC1·C      0.50345  0.21075 0.008450    .       .
Kappa·FC   0.49296  0.25739 0.027733    .       .
Kappa·C    0.50000  0.25547 0.025161    .       .
```

```
            AgreeStat_3SAS (Ver 1.0)
     INTER−RATER RELIABILITY ANALYSIS RESULTS
          Quadratic Weights (or FC Weights)
                          MODERATE_   SEVERE_
  _CATEG_NAME     NO_PAIN    PAIN      PAIN
  NO_PAIN           1.00     0.75      0.00
  MODERATE_PAIN     0.75     1.00      0.75
  SEVERE_PAIN       0.00     0.75      1.00
```

Output of SAS Program 5.1 - continued

AgreeStat_3SAS (Ver 1.0)
INTER—RATER RELIABILITY ANALYSIS RESULTS
Weighted Inter-Rater Reliability Coefficients,
Associated Standard Errors, and P-values
(Quadratic Weights)

Method	Estimate	StdErr	Pvalue
AC1	0.75229	0.11557	3.7741E-11
Kappa·F	0.74528	0.17274	.000008000
Bren.-Pred.	0.75000	0.11180	9.8517E-12
AC1·C	0.75342	0.11415	2.0515E-11
Kappa·FC	0.74286	0.17762	.000014427
Kappa·C	0.75000	0.16359	.000002274

AgreeStat_3SAS (Ver 1.0)
INTER—RATER RELIABILITY ANALYSIS RESULTS

Linear Weights (or CA Weights)

_CATEG_NAME	NO_PAIN	MODERATE_PAIN	SEVERE_PAIN
NO_PAIN	1.0	0.5	0.0
MODERATE_PAIN	0.5	1.0	0.5
SEVERE_PAIN	0.0	0.5	1.0

AgreeStat_3SAS (Ver 1.0)
INTER—RATER RELIABILITY ANALYSIS RESULTS
Weighted Inter-Rater Reliability Coefficients,
Associated Standard Errors, and P-values
(Linear Weights)

Method	Estimate	StdErr	Pvalue
AC1	0.62759	0.16209	.000053985
Kappa·F	0.61972	0.21276	.001790926
Bren.-Pred.	0.62500	0.16771	.000096971
AC1·C	0.62887	0.16101	.000046950
Kappa·FC	0.61702	0.21613	.002153143
Kappa·C	0.62500	0.20629	.001223929

5.3 The AgreeStat_3SAS Macro.

Output of SAS Program 5.1 - continued

```
           AgreeStat_3SAS (Ver 1.0)
      INTER-RATER RELIABILITY ANALYSIS RESULTS
              Radical Weights
                            MODERATE_   SEVERE_
_CATEG_NAME     NO_PAIN       PAIN       PAIN
NO_PAIN         1.00000      0.29289    0.00000
MODERATE_PAIN   0.29289      1.00000    0.29289
SEVERE_PAIN     0.00000      0.29289    1.00000
```

```
           AgreeStat_3SAS (Ver 1.0)
      INTER-RATER RELIABILITY ANALYSIS RESULTS
   Weighted Inter-Rater Reliability Coefficients,
     Associated Standard Errors, and P-values
                (Radical Weights)
Method         Estimate      StdErr       Pvalue
AC1            0.56317       0.18647     .001262987
Kappa·F        0.55555       0.23259     .008459025
Bren.-Pred.    0.56066       0.19648     .002161747
AC1·C          0.56442       0.18578     .001190543
Kappa·FC       0.55294       0.23477     .009255807
Kappa·C        0.56066       0.22836     .007041864
```

Selecting a specific group of subjects will induce an error (also called sampling error) in the agreement coefficient. This is so, because if you select another group of subjects, you will likely get a different coefficient. The StdErr is an attempt to quantify that error.

Note that the P-value establishes statistical significance with respect to the selection of subjects alone. If one wants to consider the selection of raters as another source of variation, then one should use the U_P-value instead if it is available (see Gwet (2010) for more on this).

The Weighted Analysis

When the weighted analysis is requested, the macro will compute the weighted versions of the 6 agrement coefficients under investigation using the Quadratic, Linear, and Radical weights. The first table in the weighted analysis section displays the Quadratic weights used. All diagonal elements are 1, meaning each time 2 raters choose 2 identical categories, they receive a full (agreement) credit of 1. When 2 raters select 2 adjacent categories (No-Moderate or Moderate-Severe) they receive a partial credit of 0.75. And if they select the 2 extreme categories (No and Severe pains), they are considered in total disagreement, with a 0 credit.

For each weight type, the corresponding weighted agreement coefficients along with associated standard errors, and p-values are calculated. *Note that a P-value of 3.7741E-11 is the scientific notation of a very small number, which actually represents the tiny number 0.000000000037741. That is, you need to move the decimal point to left of 3 eleven times (hence the number 11 after letter E)*

The Macro Usage: Description of Program 5.1

Program 5.1 shows an example of the SAS code you must write in order to compute the various agreement coefficients for multiple raters presented in this chapter.

Calling the AgreeStat_3SAS Macro

Line 01 indicates that "c:\advancedanalytics\myproject" is the directory that contains the AgreeStat_3SAS macro. This directory is assigned to the macro variable pdir. You are expected to provide your own directory name here.

Defining Categories *(Lines 02 through 09)*

The dataset `CategoryFile` must be provided by the user. This is typically a small dataset that lists all categories into which the rater

5.3 The AgreeStat_3SAS Macro. - 95 -

may classify a subject. Program 5.1 features a reliability experiment involving only 3 categories listed as No_Pain, Moderate_Pain, and Severe_Pain. Here are a few important things:

- ▶ Although the categories used in this example are of character type, the macro would accept numeric categories as well. These categories could well take any numeric value such 6.8, 79.51, or any integer number such as 4, 7, 9. The macro will treat all numeric-type categories as interval data for the purpose of computing weighted agreement coefficients.
- ▶ If the user opts for character-type categories, these must be written in a single word, with no space or special characters. The reason for this is that the macro uses them as variable names in a **SAS** dataset. Moreover, although character-type categories can be of any length, the macro will consider only the first 15 characters. Therefore, any 2 categories that share the same first 15 characters will be considered identical. This should not really be a major limitation with any reasonable naming conventions.

If the user does not plan to compute weighted agreement coefficients, then the order in which the character-type categories are listed in this data step is irrelevant. However,

> *if weighted coefficients are to be computed, then it is essential that the character-type categories be listed in the CategoryFile following their natural hierarchical order. In Program 5.1, if the categories had been listed as Moderate_Pain, No_Pain, and Severe_Pain, then the macro would have treated a No_Pain-Severe_Pain disagreement as a partial agreement and a Moderate_Pain-Severe_Pain disagreement as a total disagreement with a 0 weight. This would have led to wrong weighted coefficients.*

If the categories are numeric, then the order in which they are listed in the data step is irrelevant. The macro will first sort them in ascending order and weight them properly.

Some readers may prefer to create an external file containing the list of categories, in CSV, Text, or Excel format for example, and read it from **SAS**. If a csv file named CategoryNames.csv is stored in the project directory, then it may be read by replacing the lines 01 through 09 with the following:

```
01    PROC IMPORT OUT = CategoryFile
02       DATAFILE = "&ProjectDir\CategoryNames.csv"
03       DBMS = CSV REPLACE;
04       GETNAMES = YES;
05    RUN;
```

This approach has the advantage of keeping the program length fixed while offering the flexibility to use an input file of any length. Users interested in importing other data formats into **SAS** could find more information at
   http://support.sas.com/documentation/cdl/en/proc /61895/HTML/default/viewer.htm#a000332605.htm

Supplying Ratings *(Lines 10 through 20)*

This portion of the program is where the user describes the input data. Our example describes an experiment involving 3 raters named Rater1, Rater2, and Rater3, who scored 6 subjects. This is a file of raw data that lists the specific category into which each rater has classified each subject. This is one way of organizing a rating file that is similar to Table 5.1. If the ratings are supplied in this form, then the macro I wrote assumes that each variable in the input dataset is associated with a rater and related ratings. Therefore, the user should not include other variables in that dataset that do not contain ratings. This constraint is the premium to be paid for not having to supply the rating variables to the macro. I

5.3 The AgreeStat_3SAS Macro.

will discuss later in this chapter how the input data can take the form of a distribution of raters by subject and category.

Normally, one would want to codify (ideally with numbers) the category names so that lengthy names are not repeated several times in the dataset as is the case in the program. A dataset that reproduces many lengthy names can get unnecessarily big, is difficult to visualize, in addition to being error-prone. A single typographical error will lead to wrong results. Nevertheless, I have noticed in my practice that several researchers tend to create their databases using actual category names.

The treatment of category names by the macro is not case sensitive. Therefore, category names may be supplied with a mixture of lower and upper case letters. The only detail that matters is to ensure that category names in both the category and rating files match perfectly well. Otherwise, the results could be unpredictable.

Once again, users may import an external file using the IMPORT procedure of SAS as shown earlier.

Loading AgreeStat_3SAS into Your Program *(Line 21)*

The user should first load the macro's programming statement into the program containing the data. This is achieved with the single line of code:

%INCLUDE "&pdir\AgreeStat_3SAS.sas";

Users wanting to further explore the %INCLUDE statement, could find more information at:

http://support.sas.com/documentation/cdl/en/lrdict/63026/HTML/default/viewer.htm#a000214504.htm

Calling the AgreeStat_3SAS Macro *(Lines 22 through 26)*

Now that the AgreeStat_3SAS Macro has been loaded into your personal program, it must be executed as shown in Program 5.1's lines 22 through 26. This macro requires 4 parameters that I will now review:

- **InputFile =**
  This parameter takes the name of the file containing input ratings.

- **CategFile =**
  This parameter takes the name of the file containing the category names, or the numeric score values.

- **InputType =**
  This parameter tells the macro how the input ratings are organized. It accepts 2 values, which are R (lower or upper case) and S (lower or upper case). Letter R stands for "Raw", and indicates that the variables in the input file are the raters, and the records the subjects. This is how the data in Program 5.1 is organized, hence the value of the parameter. Letter S on the other hand, stands for "Summary" and indicates that variables in the input file are categories, and the records the subjects. Table 5.2 is organized this way.

  If letter S is assigned to this parameter, when the input data is of type R, my macro will not detect this error, and the results could be unpredictable.

- **Weighted =**
  This parameter accepts 2 values y for "yes" a weighted analysis is requested, and n for "no" to indicate that only the simple unweighted analysis is requested.

5.3 The AgreeStat_3SAS Macro.

Supplying Input Data in the Form of a Distribution of Raters by Subject and Category

Consider the following SAS program:

```
01  %LET pdir=c:\advancedanalytics\myproject;
02  DATA CategoryFile;
03      LENGTH CategName$ 15;
04      INPUT CategName;
05      DATALINES;
06  No_Pain
07  Moderate_Pain
08  Severe_Pain
09  ;
10  DATA RatingFile;
11      INPUT No Moderate Severe;
12      DATALINES;
13  3 0 0
14  0 1 2
15  2 1 0
16  0 0 3
17  0 1 2
18  0 3 0
19  ;
20  %INCLUDE "&pdir\AgreeStat_3SAS.sas";
21  %AgreeStat(
22      InputFile = RatingFile,
23      CategFile = CategoryFile,
24      InputType = S,
25      Weighted = y);
```

Program 5.2. Illustration of the AgreeStat_2SAS Macro

The results generated by this program are displayed in Output 5.2. One may notice that Output 5.2 is a subset of Output 5.1. This is due to the fact that the input data supplied to Program 5.2 (see lines 10 through 19) is more restrictive than that of Program 5.1; making it impossible to compute to implement the agreement coefficients based on Conger's generalization technique[7]. Note that without the raw data of Program 5.1, it is also impossible to compute the unconditional standard error, and the associated U_Pvalue that account for the variation in agreement coefficient due to the selection of raters.

There are 2 main differences between Programs 5.1 and 5.2.

▶ The Input data (Lines 10 through 19 in Program 5.2)
One should note that the variables used in that data step need not match the list of variables in the CategoryFile. Any variable names can be used here indeed. Nevertheless, you need to be cautious here for the following reason:

> *If weighted agreement coefficients are to be requested, then it is essential that the variables in the input dataset (i.e. in line 11) appear in the same order they are listed in CategoryFile. That is variable No will match the No_Pain category, Moderate will match the Moderate_Pain category, and Severe will match the Severe_Pain category. If the order in which input variables are stored in the input file is inconsistent with CategoryFile, then the weighted coefficients will be wrong, although simple coefficients will still be correct.*

▶ The InputType Paramater (Line 24 in Program 5.2)
Because the input data type has changed, this parameter must reflect that. Otherwise, the macro is likely to generate quite a bit of errors.

---

[7]These agreement coefficients are Kappa.C, Kappa.FC, and AC1.C

5.3 The AgreeStat_3SAS Macro.

Output 5.2. Output of SAS Program 5.2

```
           AgreeStat_3SAS (Ver 1.0)
    INTER-RATER RELIABILITY ANALYSIS RESULTS

Unweighted Inter-Rater Reliability Coefficients,
   Associated Standard Errors, and P-values
    Method     Estimate  StdErr    Pvalue
    AC1        0.50230   0.22093   0.011497
    Kappa·F    0.49533   0.23157   0.016219
    Br.-Pred.  0.50000   0.22361   0.012674
```

```
           AgreeStat_3SAS (Ver 1.0)
    INTER-RATER RELIABILITY ANALYSIS RESULTS

       Quadratic Weights (or FC Weights)
                            MODERATE_  SEVERE_
  _CATEG_NAME    NO_PAIN     PAIN       PAIN
  NO_PAIN        1.00        0.75       0.00
  MODERATE_PAIN  0.75        1.00       0.75
  SEVERE_PAIN    0.00        0.75       1.00
```

```
           AgreeStat_3SAS (Ver 1.0)
    INTER-RATER RELIABILITY ANALYSIS RESULTS
 Weighted Inter-Rater Reliability Coefficients,
    Associated Standard Errors, and P-values
                 (Quadratic Weights)
    Method      Estimate  StdErr      Pvalue
    AC1         0.75229   0.11557   3.7741E-11
    Kappa·F     0.74528   0.17274   .000008000
    Bren.-Pred. 0.75000   0.11180   9.8517E-12
```

Output of SAS Program 5.2 - continued

AgreeStat_3SAS (Ver 1.0)
INTER—RATER RELIABILITY ANALYSIS RESULTS
Linear Weights (or CA Weights)

_CATEG_NAME	NO_PAIN	MODERATE_PAIN	SEVERE_PAIN
NO_PAIN	1.0	0.5	0.0
MODERATE_PAIN	0.5	1.0	0.5
SEVERE_PAIN	0.0	0.5	1.0

AgreeStat_3SAS (Ver 1.0)
INTER—RATER RELIABILITY ANALYSIS RESULTS
Weighted Inter-Rater Reliability Coefficients,
Associated Standard Errors, and P-values
(Linear Weights)

Method	Estimate	StdErr	Pvalue
AC1	0.62759	0.16209	.000053985
Kappa·F	0.61972	0.21276	.001790926
Bren.-Pred.	0.62500	0.16771	.000096971

5.3 The AgreeStat_3SAS Macro.

Output of SAS Program 5.2 - continued

AgreeStat_3SAS (Ver 1.0)
INTER—RATER RELIABILITY ANALYSIS RESULTS
Radical Weights

_CATEG_NAME	NO_PAIN	MODERATE_ PAIN	SEVERE_ PAIN
NO_PAIN	1.00000	0.29289	0.00000
MODERATE_PAIN	0.29289	1.00000	0.29289
SEVERE_PAIN	0.00000	0.29289	1.00000

AgreeStat_3SAS (Ver 1.0)
INTER—RATER RELIABILITY ANALYSIS RESULTS
Weighted Inter-Rater Reliability Coefficients,
Associated Standard Errors, and P-values
(Radical Weights)

Method	Estimate	StdErr	Pvalue
AC1	0.56317	0.18647	.001262987
Kappa·F	0.55555	0.23259	.008459025
Bren.-Pred.	0.56066	0.19648	.002161747

Weighting Issues

I discussed the issue of weights in chapter 4 when the number of raters was limited to 2. Increasing the number of raters to 3 or more, does not really change the situation fundamentally. The weights as previously discussed, solely depend upon the categories. When these categories are of character type, then all the weights are calculated using sequential integers 1, 2, until the number of categories. When the categories are numeric, then these numbers determine the weight values. Using quadratic weights with numeric categories will yield the agreement coefficients discussed by Gwet (2010) for ordinal and interval data. The weighted AC_1 with quadratic weights yields the AC_2 coefficient of Gwet (2010).

The current version of the **AgreeStat_3SAS** macro does not offer the user the option of supplying custom weights. This limitation will likely be corrected in the next version of the macro. I also caution users about the overuse of weights when computing agreement coefficients. In fact, these weights should be decided upon before the actual calculations are performed. It is unethical to play with a variety of weight values for the sole purpose of obtaining a high agreement coefficient to support the quality of our research data.

CHAPTER 6

Rater Agreement with SAS Enterprise Guide

6.1 Introduction

This chapter is devoted to SAS Enterprise Guide (EG) users. I had to modify the AgreeStat_2SAS and AgreeStat_3SAS macros slightly to adapt them to the Enterprise Guide environment. These changes led to the new names AgreeStat_2SASEG and AgreeStat_3SASEG, where the letters E and G stand for Enterprise Guide. In order to simplify as much as possible the use of these macros to non-programmer EG users, I have created two EG project files named AgreeStat_2EG.egp, and AgreeStat_3EG.egp that can be downloaded and modified. These EG projects use some test input datasets that could be downloaded as well for the purpose of testing the project file. All these files can be downloaded at:

http://www.agreestat.com/agreestat.html

I will show you step by step in the next two sections how these 2 project files can be used to compute inter-rater reliability coefficients for 2 raters, and for the more general situation of 3 raters or more. You will only need to modify the input data files before re-running the whole project file.

6.2 Agreement Coefficients for 2 Raters

Testing the AgreeStat_2EG.egp Project File

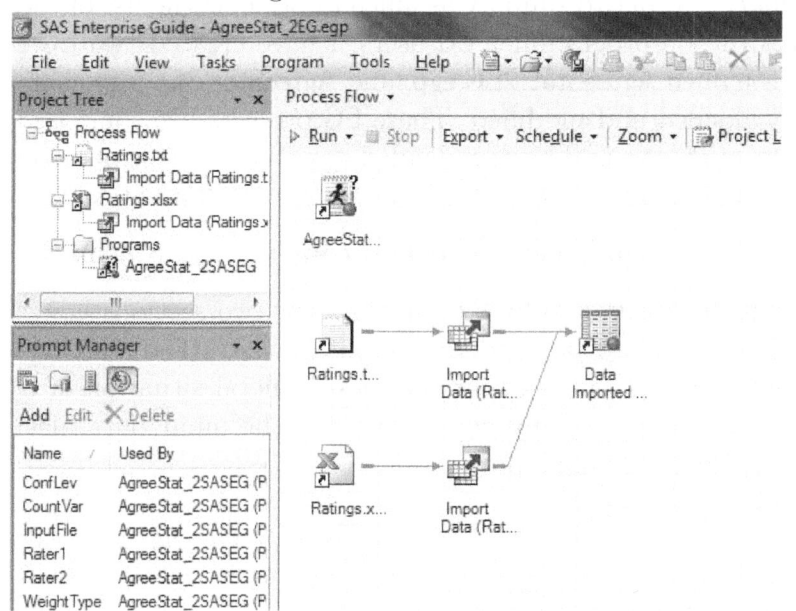

The Enterprise Guide project file AgreeStat_2EG.egp is what you will need to compute the extent of agreement among 2 raters. To test it use the following 8-step procedure:

1. **Downloading AgreeStat_2EG.egp**

    Download the EG project file AgreeStat_2EG.egp as well as the macro `AgreeStat_2SASEG.sas`, and the 2 files Ratings.txt and Ratings.xlsx using the URL shown above. If the user does not have MS Excel then the Excel file may be ignored.

2. **Opening the AgreeStat_2EG.egp File**

    After opening this EG project file, you should see a process flow as shown in Figure 6.1.

Figure 6.1. The Enterprise Guide Project File AgreeStat_2EG.egp

6.2 Agreement Coefficients for 2 Raters.   - 107 -

At this stage if one of the datasets will not be used, it could simply be deleted. To delete a file, right-click on it in the project tree, and select "Delete." Both the text and the Excel files are not needed. Only one of them is, and both are used here only to illustrate the 2 possibilities.

3 **Establishing Links**

Although the process flow looks fine at first glance, Enterprise Guide may still not be able to access neither the SAS macro program, nor the datasets. These links must be established for each of the 3 files of the "Project Tree," which are Ratings.txt, Ratings.xls, and AgreeStat_2SASEG. For the Ratings.txt file for example, proceed as follows:

(1) Right-click on Ratings.txt,

(2) Select "Properties,"

(3) Click on "Change,"

(4) Navigate to the location where you stored the Ratings.txt file and click on it. **This procedure must be repeated for the other 2 files.**

Note that you may use this same procedure to replace the files I put there with your own files.

4 **Execute AgreeStat_2EG.egp**

Run the project from the File menu for instance (this could be done using other buttons as well). You will then be presented with the menu shown in Figure 6.2. Fill in all the fields as shown in the figure. The first menu item is the input file containing the ratings. **Type in Ratings**. Alternatively, you could type in any other file name you may have used in place of Ratings. The second and third menu items are the names of the 2 variables in the input dataset that describe both raters' scores (in this case **Judge1** and **Judge2**)[1]. The 4th item

---
[1] Look at figures 6.5 and 6.6 to see what the input file looks like.

"Name of the Frequency Variable" is optional. If not specified, the program will add a variable named **Count** to the input dataset, and will assign 1 to it, indicating that each record only represents a single subject. In our example, the variable **Count** must be specified. The remaining 2 items are the choice of weights (for computing weighted agreement coefficients), and the confidence level for calculating confidence intervals. I discussed the weighting and confidence level issues in previous chapters.

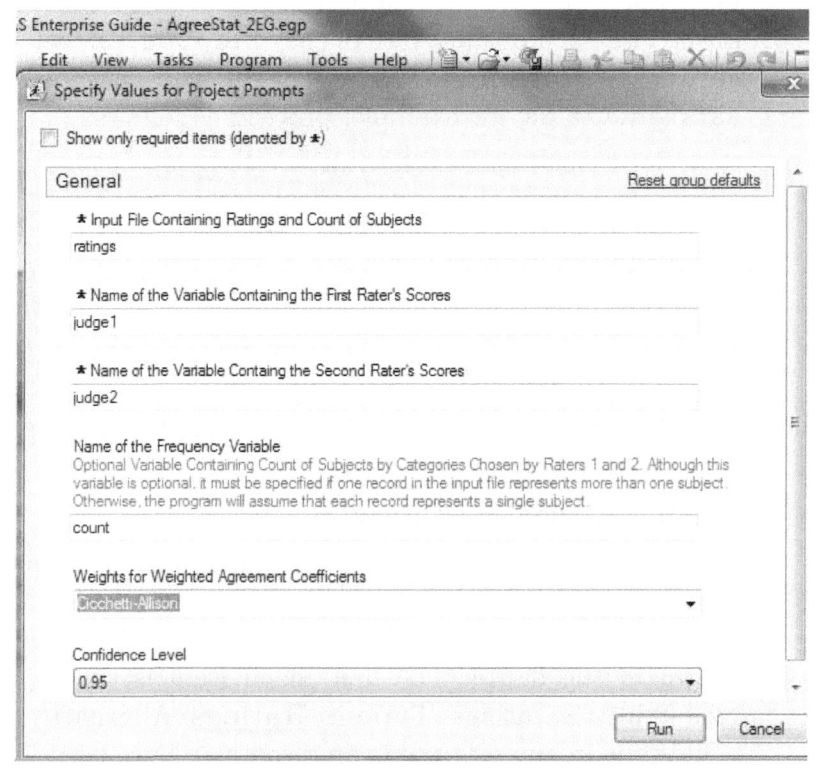

Figure 6.2. AgreeStat_2EG.egp's Selection Menu

After filling in all the fields, **Click the Run button** to execute the program.

6.2 **Agreement Coefficients for 2 Raters.** **- 109 -**

5 **Reviewing AgreeStat_2EG's Output**

After executing the program in step 4, you may view the updated process flow by clicking on "Process Flow" on the top horizontal menu. Figure 6.3 shows a portion of that process flow after execution. You can see the presence of the node "SAS Report - AgreeStat_2SASEG," which contains the results. **Click that node** to see the results, a portion of which is shown in Figure 6.4.

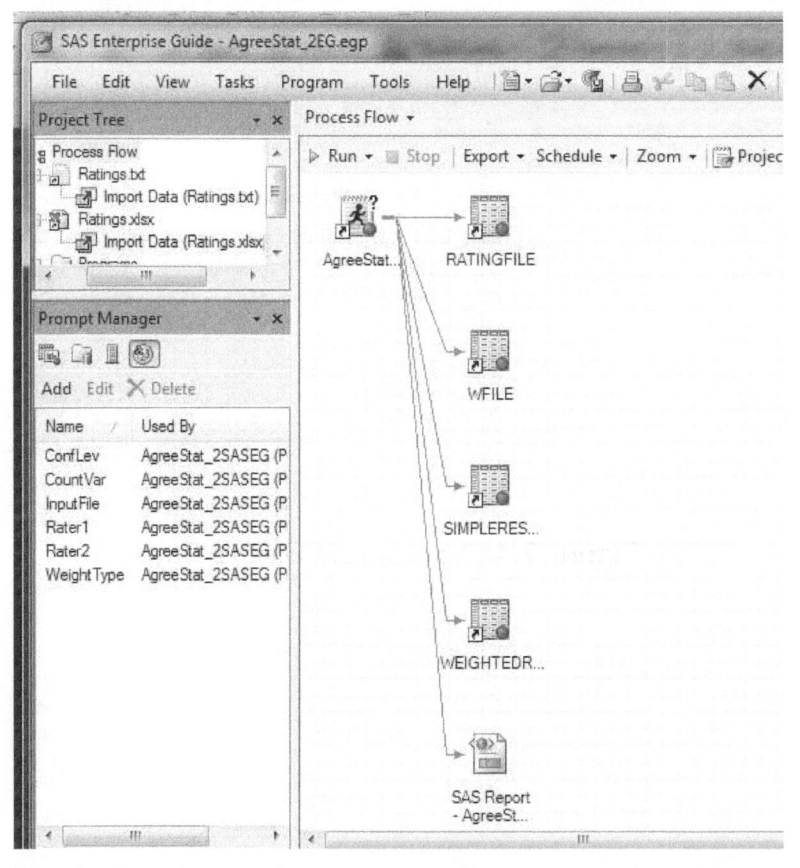

Figure 6.3. Process Flow After the Execution of AgreeStat_2EG.egp

Chapter 6. *Rater Agreement with Enterprise Guide*

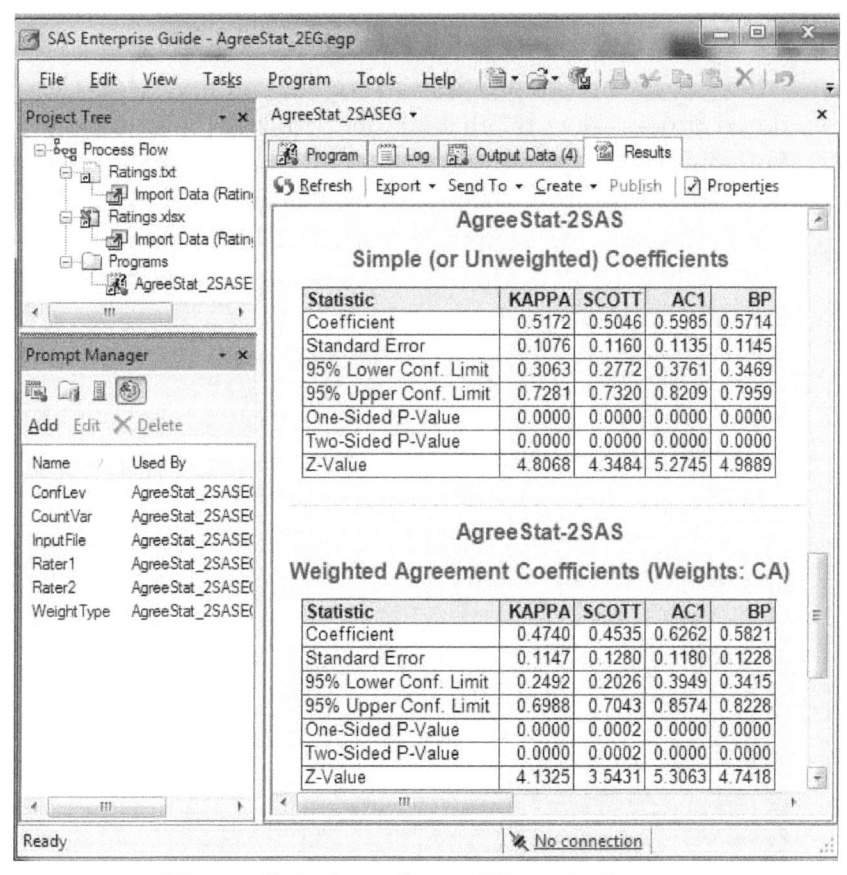

Figure 6.4. AgreeStat_2EG.egp's Output

6.2 Agreement Coefficients for 2 Raters.

Modifying the Input Files

So far, I have shown the steps for testing the **AgreeStat_2EG.egp** project file. Soon or later you will need to use your own files if you want to go beyond what I have done, and solve a problem that is important to you. If you are already familiar with **Enterprise Guide**, then you may well skip this section altogether, and create your own **SAS** dataset before running the program. If you are not yet familiar with **EG**, I offer 2 options, which involves using Microsoft Excel, and a text editor like Notepad.

USING MS EXCEL

Figure 6.5 shows the Excel file I used with the **AgreeStat_2EG** project. You may well simply open it and replace my data with yours and rerun the program as it is. This would definitely be the simplest way. But a way that requires that you have access to Microsoft Office. Excel has another advantage, which is that **SAS Enterprise Guide** already knows so well how to convert an Excel file to a **SAS** data that it does it automatically as soon as you open the Excel file from **EG**.

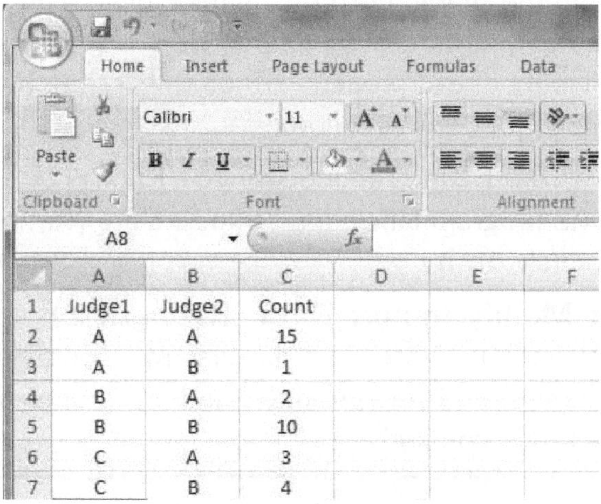

	A	B	C
1	Judge1	Judge2	Count
2	A	A	15
3	A	B	1
4	B	A	2
5	B	B	10
6	C	A	3
7	C	B	4

Figure 6.5. Ratings.xls File

USING NOTEPAD

If you do not have access to MS Office, no panic. In fact any text editor will do the job just fine. My preferred text editor for this task is Notepad, which comes with almost all new PCs these days running MS Windows. Figure 6.6 shows the text file that I used in this project. Again, without modifying the file or variable names, you could well simply replace my data with yours and run the program.

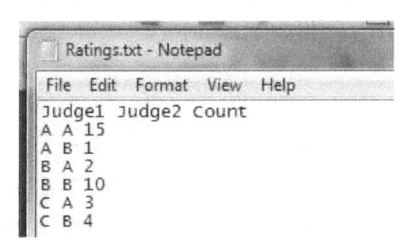

```
Ratings.txt - Notepad
File  Edit  Format  View  Help
Judge1 Judge2 Count
A A 15
A B 1
B A 2
B B 10
C A 3
C B 4
```

Figure 6.6. Ratings.txt File

Warning:
When modifying my input files and if you decide to change the data type from the character to the numeric type, then I strongly recommend that you modify the import data task to insure that the data will still be read properly. To do this for the Ratings.txt file for example, proceed as follows:

- ▶ In the Project Tree, and below the Ratings.txt file, is a little SAS program labeled as "Import Data (Ratings.txt)". **Right-click on it.**

- ▶ **Select Modify Import Data (Ratings.txt),** and follow the instructions. Note that the first row contains the variable names, and not the data, and the proper box must be checked accordingly.

6.3 Agreement Coefficients for 3 Raters or More

Testing the AgreeStat_3EG.egp Project File

In this section, I will explain how you can compute the extent of agreement among 3 raters or more using the **AgreeStat_3EG.egp** project file. If you read the previous section, you should already be familiar with the way my programs operate. Therefore my instructions in this section will be more concise. Once again, I suggest to test the EG project file **AgreeStat_3EG.egp** along with the test data that come with it before you run the program with your own data. Unless of course, you already know what you are doing. Here are the steps for testing this program:

1. **Downloading AgreeStat_3EG.egp**

   Download the EG project file **AgreeStat_3EG.egp** as well as the macro `AgreeStat_3EG.sas`, and the 4 files `Ratings3More.txt`, `CategData3More.txt` for the text files, then `Ratings3More.xlsx` and `CategData3More.xlsx` for Excel[2]. All of these files can be downloaded at:

   http://www.agreestat.com/agreestat.html

   `CategData3More.txt` only contains the list of category names, while `Ratings3More.txt` contains the actual ratings. You may want to read chapter 5 for a more detailed discussion on the role these files play.

2. **Opening the AgreeStat_3EG.egp File**

   After opening this EG project file, you should see a process flow as shown in Figure 6.7. If the Excel files are not going to be used, then **Right-click successively on each file's node and select Delete** to delete them. The text files alone will allow the program to operate properly.

---

[2]Users with no access to Excel may simply ignore the Excel files.

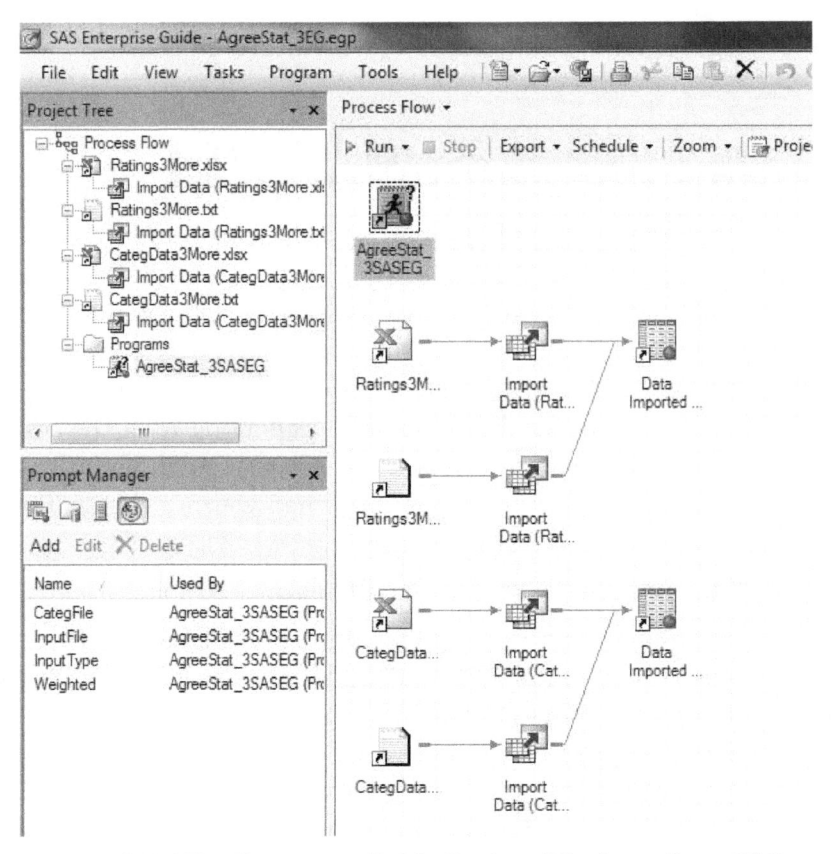

Figure 6.7. The Enterprise Guide Project File AgreeStat_3EG.egp

③ Establishing Links

Establish the links between the file names in Enterprise Guide and their physical location as discussed in the previous section.

6.3. Agreement Coefficients for 3 Raters or More.

④ Execute AgreeStat_3EG.egp

Run the project from the File menu for instance. You will then be presented with the menu shown in Figure 6.8. Fill in the fields as shown on the figure. The first menu item is the name of the input file containing the ratings. **Type in ratings3more.** The second menu item is the name of the file containing the list of category names. **Type in categdata3more.**

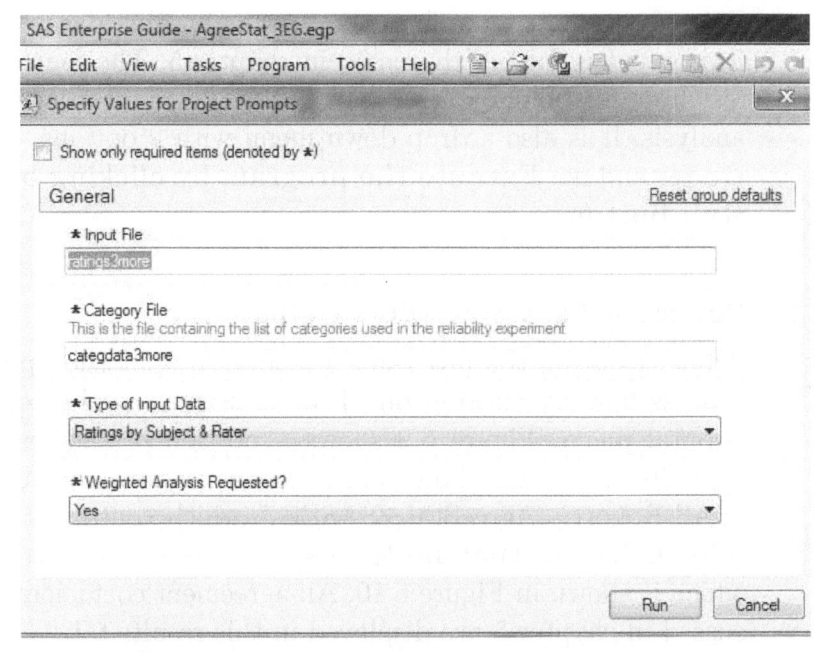

Figure 6.8. AgreeStat_3EG.egp's Selection Menu

The categdata3more file must list categories sorted following their natural hierarchical order if any. That is if the categories are No, Moderate, and Severe, then they should be listed in that order for the program to know that Severe is ranked higher than moderate, which in turn is ranked higher than No. You may want to read chapter 3 to learn everything about these 2 files. Note that these file names refer to the **SAS** datasets.

The third menu item is the "Type of Input Data." This is a drop-down menu with 2 options, which are "Ratings by Subject & Rater," and "Count of Raters by Subject and Category." In this example it is the first option that must be selected as the input dataset Ratings3More lists the ratings by subject (each record or row represents a subject) and by Rater represented by the columns. Again this dataset and its content is further discussed in chapter 5. The fourth and last item is about requesting or not requesting the weighted analysis. It is also a drop-down menu with 2 options, which are yes and no. **Execute the program by clicking on the Run button.**

5 Reviewing AgreeStat_2EG's Output

After executing the program in step 4, look at the updated process flow by clicking on "Process Flow" on the top horizontal menu. Figure 6.9 shows a portion of that process flow after execution. You can notice the presence of the node "SAS Report - AgreeStat_3SASEG," which contains the results. **Click on that node** to see the results, a portion of which is shown in Figure 6.10. All agreement coefficients discussed in chapter 5 are displayed in this results tab.

6.3. Agreement Coefficients for 3 Raters or More.

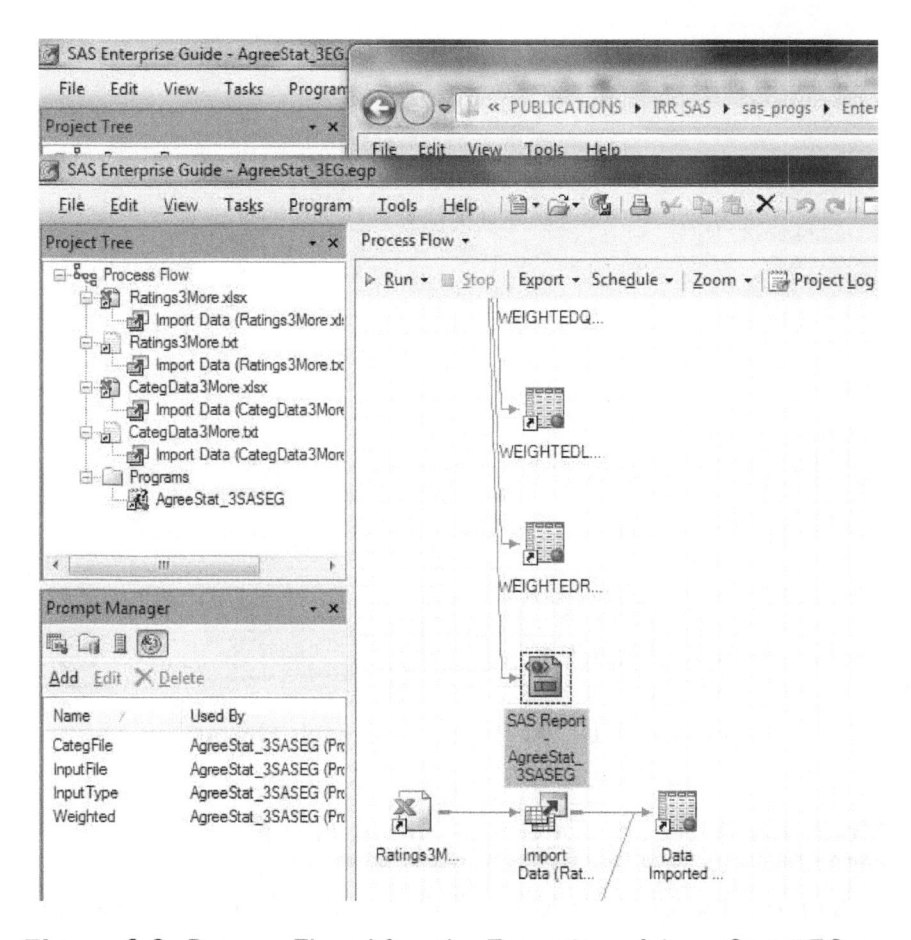

Figure 6.9. Process Flow After the Execution of AgreeStat_3EG.egp

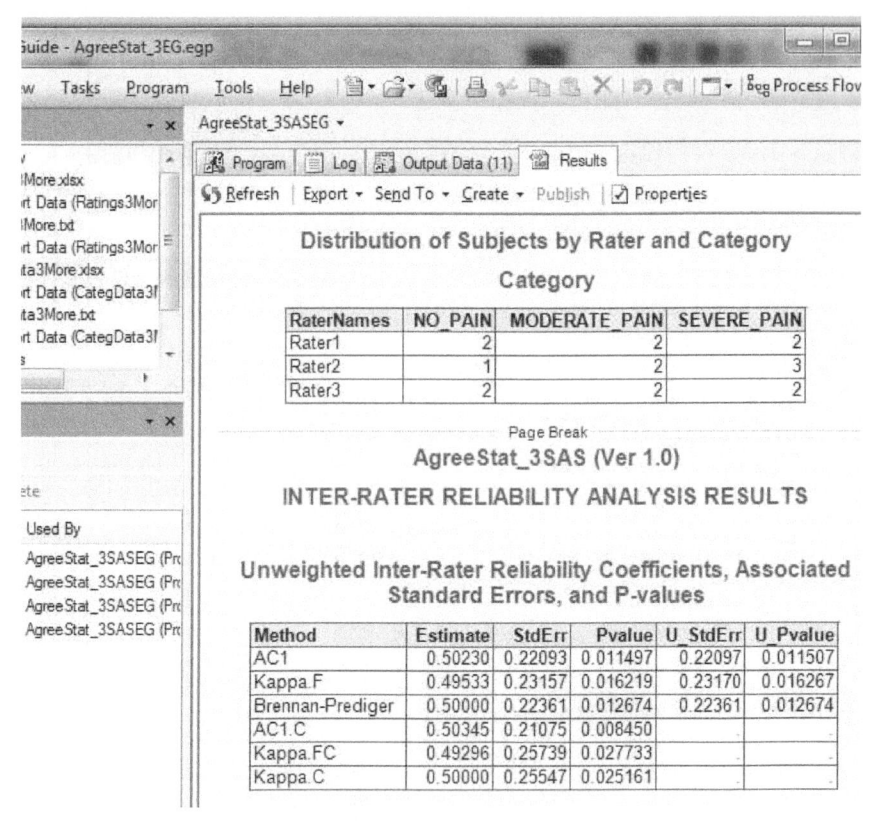

Figure 6.10. AgreeStat_3EG.egp's Output

Modifying AgreeStat_3EG.egp's Input Files

As indicated in the previous section, you will need to use your own data at some point to resolve a problem. You may decide to use Excel or a text editor such as Notepad. No need to use both though as I did to illustrate the 2 possibilities. Actually MS Access could be used as well. Once you decide what option you choose, I recommend that you delete the nodes from the process flow that will not be used.

In case you decide to use Excel or Notepad, simply modify the

6.3. Agreement Coefficients for 3 Raters or More.

file you already have by replacing my data and eventually my variable names with your own information before re-running the project file as it is. Once again, I will never insist enough on this, if you replace my data with your own data, which are of numeric type then you have to modify the Import Data task as explained in the previous section.

The 4 files used in this project are shown in Figures 6.11 (for Excel) and 6.12 (for Notepad). You may simply modify them right from where they are, eventually modify the Import Data task, then run the project file to get your results. This is as simple as it sounds.

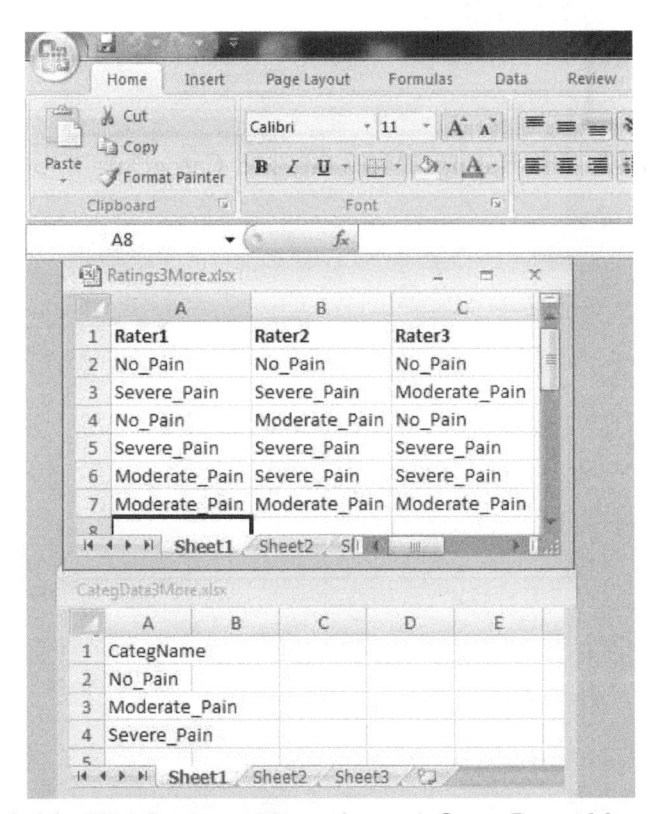

Figure 6.11. The Ratings3More.xls, and CategData3More.xls Files

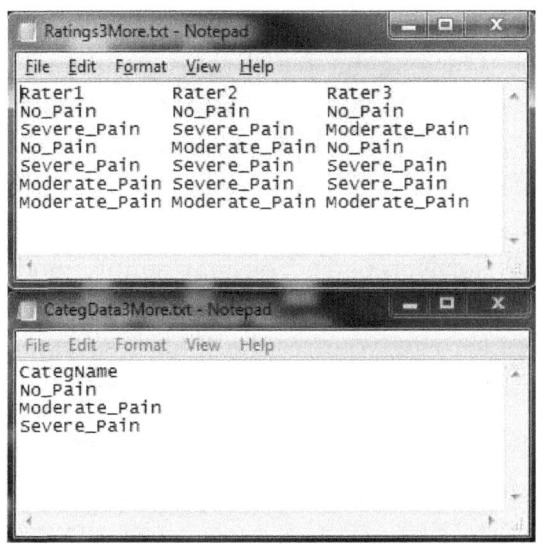

```
Ratings3More.txt - Notepad
File  Edit  Format  View  Help
Rater1          Rater2          Rater3
No_Pain         No_Pain         No_Pain
Severe_Pain     Severe_Pain     Moderate_Pain
No_Pain         Moderate_Pain   No_Pain
Severe_Pain     Severe_Pain     Severe_Pain
Moderate_Pain   Severe_Pain     Severe_Pain
Moderate_Pain   Moderate_Pain   Moderate_Pain
```

```
CategData3More.txt - Notepad
File  Edit  Format  View  Help
CategName
No_Pain
Moderate_Pain
Severe_Pain
```

Figure 6.12. The Ratings3More.txt, and CategData3More.txt Files

CHAPTER 7
Concluding Remarks

I wrote this book to assist researchers and students who may want to use the SAS system to compute inter-rater reliability coefficients for nominal, ordinal and interval data. Not long ago, I published a methodology book entitled "Handbook of Inter-Rater Reliability: The Definite Guide to Measuring the Extent of Agreement Among Multiple Raters." The current book can be seen as a SAS companion to the methodology book. While the methodology book discusses the merits of the different techniques at length, this SAS companion aims at assisting practitioners with their SAS-based production work. It would not have been possible to add this material to the methodology book as a number of practitioners do not use SAS. Hence the release of this book.

The AGREE option provided in the FREQ procedure is a welcome addition to the SAS system. It allows SAS users to easily obtain simple and weighted Kappa statistics using an already widely-used procedure. Standard errors and P-values can also be obtained for researchers interested in statistical inference. The FREQ procedure also provides the option to compute exact P-values using either the network algorithm or the Monte-Carlo simulation approach.

The implementation of Kappa in SAS is unfortunately limited to two raters and does not work when the data carries the unbalanced-table problem, or produces erroneous results on balanced tables that have the diagonal problem. I discussed these issues at length in chapter 3, and offered a solution for PC SAS users with the SAS macro program called AgreeStat_2SAS.sas; and another solution for SAS Enterprise Guide users called AgreeStat_2EG.egp. With these solutions, the user will always get correct results without having to worry about the unbalanced-table and diagonal problems. The user can also supply character as well as numeric type data.

The only constraint with the use of AgreeStat_2SAS is for you to use a suitable category naming convention. If the weighted analysis is of interest then the categories must be named so that their alphabetical order matches their natural hierarchical order as well. Suppose you need to use categories such as VeryDissatisfied, Dissatisfied, Satisfied, and VerySatisfied. Their alphabetical order will be Dissatisfied, Satisfied, VeryDissatisfied, and VerySatisfied. When computing weighted coefficients, SAS will treat VeryDissatisfied as being closer to VerySatisfied than to Satisfied. This will lead to wrong results. If you have to use these categories, I suggest to re-label them as 01VeryDissatisfied, 02Dissatisfied, 03Satisfied, and 04VerySatisfied. This will solve the problem.

For the more general situation of 3 raters or more, the SAS Institute's support group offers the macro program `magree.sas`, which implements the generalized Kappa statistics proposed by Fleiss (1971). In addition to producing the overall Kappa, this macro can compute Kappa coefficients conditionally on a specific response category. These conditional statistics allow researchers to evaluate the propensity of raters to agree on a specific category. Practitioners interested only in Fleiss' generalized Kappa may use this macro, which can be downloaded from SAS Institute's website. I must say that I am personally skeptical about the validity of the equations used to compute the standard errors in this macro. I know very well that these equations came from the book Fleiss et al. (2003). Gwet (2008a) presents the correct equation for the standard error of Fleiss generalized Kappa.

For this general situation of 3 raters or more, I proposed ano-

6.3. Agreement Coefficients for 3 Raters or More.

ther SAS macro called AgreeStat_3SAS.sas for PC SAS users, and a SAS Enterprise Guide project file AgreeStat_3EG.egp for Enterprise Guide users. Both can be downloaded at:

www.agreestat.com/agreestat.html

My solution implements Fleiss' generalized Kappa, the Brennan-Prediger coefficient, Conger's generalized Kappa when the data is available, Gwet's AC_1 as well as as their weighted versions.

You will note that the solutions I offer are all based on the SAS/IML language also known as Proc IML. You will need this licence to be able to use these solutions. Similar solutions using Base SAS only will require considerable programming efforts to implement all agreement coefficients and associated standard errors that I have discussed.

Bibliography

[1] Bennett et al. (1954). Communications through limited response questioning. *Public Opinion Quarterly*, **18**, 303-308.

[2] Brennan, R. L., and Prediger, D. J. (1981). Coefficient Kappa: some uses, misuses, and alternatives. *Educational and Psychological Measurement*, **41**, 687-699.

[3] Cicchetti, D.V. and Allison, T. (1971). A new procedure for assessing reliability of scoring EEG sleep recordings. *American Journal of EEG Technology*, **11**, 101-109.

[4] Cicchetti, D.V. and Feinstein, A.R. (1990). High agreement but low Kappa : II. Resolving the paradoxes. *Journal of Clinical Epidemiology*, **43**, 551-558.

[5] Conger, A.J. (1980). Integration and Generalization of Kappas for Multiple Raters. *Psychological Bulletin*, **88**, 322-328.

[6] Cohen, J. (1960). A coefficient of agreement for nominal scales. *Educational and Psychological Measurement*, **20**, 37-46.

[7] Cohen, J. (1968). Weighted kappa : Nominal scale agreement with provision for scaled disagreement or partial credit. *Psychological Bulletin*, **70**, 213-220.

[8] Crewson, P.E. (2001). A Correction for Unbalanced Kappa Tables.
http://www2.sas.com/proceedings/sugi26/p194-26.pdf

[9] Fleiss, J. L. (1971). Measuring nominal scale agreement among many raters., *Psychological Bulletin*, **76**, 378-382.

[10] Fleiss, J.L., Cohen J. (1973). The equivalence of weighted kappa and the intraclass correlation coefficient as measures of reliability. *Educational and Psychological Measurement*, **33**, 613-619.

[11] Fleiss, J.L., Cohen J., and Everitt, B.S. (1969). Large sample standard errors of kappa and weighted kappa. *Psychological Bulletin*, **72**, 323-327.

[12] Fleiss, J.L., Levin, B., Paik, M.C. (2003). *Statistical Methods for Rates and Proportions (3rd edition)*. Wiley-Interscience.

[13] Gwet, K.L. (2008a). Computing inter-rater reliability and its variance in the presence of high agreement. *British Journal of Mathematical and Statistical Psychology*, **61**, 29-48.

[14] Gwet, K.L. (2008b). Variance estimation of nominal-scale inter-rater reliability with random selection of raters. *Psychometrika*, **73**, 407-430.

[15] Gwet, K.L. (2010). *Handbook of Inter-Rater Reliability (2nd edition)*. Advanced Analytics, LLC.

[16] Kilpikoski, S., Airaksinen, O., KanKaanpaää, M. et al. (2002). Interexaminer reliability of low back pain assessment using the McKenzie method. *Spine*, **27**, E207-214.

[17] Light, R.J. (1971). Measures of response agreement for qualitative data: some generalizations and alternatives, *Psychological Bulletin*, **76**, 365-377.

[18] Liu, H. and Hays, R.D. (1999). Measurement of Interrater Agreement : A SAS/IML Macro Kappa Procedure for Handling Incomplete Data. *Proceedings of the Twenty-Fourth Annual SAS Users Group International Conference, April 11-14, 1999*, 1620-1625.

[19] Mehta, C.R. and Patel, N.R. (1983). A Network Algorithm for Performing Fisher's Exact Test in $r \times c$ Contingency Tables. *Journal of the American Statistical Association*, **78**, 427-434.

[20] Scott, W. A. (1955). Reliability of Content Analysis : The Case of Nominal Scale Coding. *Public Opinion Quarterly*, **XIX**, 321-325.

[21] Siegel, S. and Castellan Jr., N.J. (1988). *Nonparametric Statistics for the Behavioral Sciences (2nd ed.)*. New York : McGraw-Hill. p. 266.

[22] Stein, C.R., Devore Jr., R.B., and Wojcik, B.E. (2005).Calculation of the Kappa Statistic for Inter-rater Reliability: The Case Where Raters Can Select Multiple Responses from a Large Number of Categories. *Proceedings of the Thirtieth Annual SAS Users Group International Conference, April 10-13, 2005.*

[23] Zwick, R. (1988). Another look at Interrater Agreement. *Psychological Bulletin*, **103**, 374-378.

Author Index

Airaksinen, O., 10
Albert, R., 14, 16
Allison, T., 3, 14, 20, 45-47, 58, 88

Bennett, E. M., 14, 16
Brennan, R. L., 4, 14, 18, 26, 30, 45, 56, 59, 85, 123

Catellan Jr., N.J., 78
Cicchetti, D. V., 3, 14, 20, 45-47, 58, 88
Cohen, J., 2, 3, 12, 13, 16, 19-21, 29, 45, 48, 56, 58, 61, 65 81, 83, 86, 88
Conger, A. J., 4, 29, 83-87, 100, 123
Crewson, P.E. 39

Devore Jr., R.B. 39

Everitt, B., 61, 65

Fleiss, J. L. 2-4, 20, 21, 29, 42, 45, 48, 58, 61, 65, 77, 78, 81, 83, 85, 87, 88, 122, 123

Goldstein, A. C., 14, 16
Gwet, K. L., ix, 4, 12, 14-16, 27, 29, 30, 33, 34, 45, 56, 58, 59, 61, 65, 85, 86, 88, 90, 93, 104, 122, 123

Hays, R.D., 39

Kankaanpää, M., 10
Kilpikoski, S., 10

Levin, B., 42
Light, R. J., 29
Liu, H., 39

Mehta, C.R., 71

Paik, M.C., 42
Patel, N.R., 71
Prediger, D. J., 4, 14, 18, 26, 30, 45, 56, 59, 85, 123

Scott, W. A., 13, 16-18, 25, 30, 45, 56, 59, 63, 64, 87
Siegel, S., 78
Stein, C.R., 39

Wojcik, B.E., 39

Zwick, R., 33

Subject Index

A
AC$_1$, 4, 14-16, 18, 27, 30, 85-88, 90
AC1.C, 90
AC1.FC, 90
AC1C, 86
AC1C Coefficient, 86-87
AC$_2$, 65, 104,
AGREE option, 4-5, 12, 31, 35, 45, 47, 52-53, 66, 70-71, 73, 121
Agreement
 Credit, 8, 20, 94
 Probability, 11-12, 14, 16-17, 22, 80-81, 84
 Table, 5-6, 42
AgreeStat_2EG.egp
 Project File, 105-111
AgreeStat_2SAS
 Macro, 12, 45, 56-58, 89, 99, 105
 Macro Example, 57
 Macro Output, 62-64
AgreeStat_3EG.egp
 Project File, 105, 113-115, 1178
AgreeStat_3SAS
 Macro, 77, 88-89, 94, 97-98, 105
 Macro Example, 89
 Macro Output, 91
ALPHA Option, 73
AP Probability
 for 2 Raters, 11-14, 16-18
AP Probability for 3 Raters or More, 80-81, 83, 85-87
ASE, 33-34, 36, 43-44, 46, 54-55, 62-63, 69-70, 72, 74-75
Asymptotic Standard Error, 33, 70-71

B
Balanced Table 6, 38-39, 121
Bennett's S Coefficient, 14
BP Coefficient, 18, 26, 59, 63-64, 85
Brennan-Prediger Coefficient, 4, 14, 18, 26, 30, 45, 56, 59, 85, 123
 Generalized Coefficient, 85, 123

C
CA Weights, 20-21, 23, 45-51, 58, 89, 92, 102
 for 3 Raters or More, 81, 82, 83,

Subject Index

85
CAP Probability
   for 2 Raters, 11-15, 17-18, 80-88
   AC_1, 15, 18, 27
   Brennan-Prediger, 18, 26, 85
   Kappa, 11-12, 17
   Scott's Pi, 13, 17
   S-index, 14, 18
CAP.C, 83-86
CAP.F, 81-87
CAP.FC, 87-88
CAP.G, 85-86
CAP.GC, 86-87
Category-Level
   Chance-Agreement Probability, 82
   Classification Probability, 82
CCAP, 82
CCP, 82
Cell-Level Agreement, 80
Chance-Agreement Probability, 11-12, 14, 17-18, 22, 80-82, 84, 86
Chi-Square, 6, 44, 63
Classification Propensity, 12-13, 15, 122
Conditional Analysis, 78, 122
Confidence Interval, 2, 33, 56, 59-60, 108
Confidence Level, 59-60, 73, 108
Confidence Limits, 34, 36, 43-44, 46, 54-55, 62-63, 73
Conger's Generalized Kappa, 83
Contingency Table, 4-6, 10, 30-32, 39, 50-53, 71

Custom Weights, 10, 45, 58, 104

D

Diagonal Problem, x, 4-6, 37-38, 42, 45, 56, 59, 121-122
   Solution, 38, 42, 45
Disagreements, 3, 7-8, 10, 15-16, 19, 46, 50-51
   Serious, 3, 7-8, 10, 15, 19, 46, 50-51
Dummy Ratings, 39, 41

E

Exact P-values, 70
EXACT Statement, 70-71, 73
   AGREE option, 71, 73

F

FC Weights, 20-21, 23, 26, 45-51, 57-58, 62-65, 89, 91, 101
Fleiss Generalized Kappa, 77, 81, 83, 122-123
   Chance-Agreement Probability, 81
FORMAT Procedure, 51, 53
FREQ Procedure, x-xi, 2-6, 11, 29-32, 34-35, 37, 45

G

Gwet's AC_1, 4, 14-16, 18, 27, 30, 45, 56, 59, 63-64, 85-86
Gwet's AC_2, 65, 104

Subject index

I
%INCLUDE Statement, 57, 89, 97, 99
IMPORT Procedure, 96-97, 112, 119
Initial Seed, 73-75
Inter-Rater Reliability, ix, x, 10, 14-15, 31, 59-60, 78

J
Jackknife, 61, 65

K
Kappa Coefficient, 11, 13, 15, 17
KAPPA option, 4-5
Kappa.C, 85, 90
KAPPA.F, 81, 83
Kappa.FC, 87

L
Linear Weights, 20-21, 45, 89, 92, 94, 102

M
Magree Macro, 2, 77-78, 122
Marginal Homogeneity, 32-33
MC Option, 71, 73
Monte-Carlo, 71, 73-75, 121
  P-Value, 73
Multiple Raters, 77-78

O
Ordinal Data Problem, x, 7
  Solution, 50-54

P
Partial Agreement, 15, 19, 94-95
Pi Coefficient, 13-14, 16-18, 25, 30, 45, 56, 59, 87
P-value, 33, 60, 63-64, 70, 73, 90, 93
  Exact, 70
  One-Sided, 60, 63-64, 73
  Two-Sided, 60, 63-64
  Interpretation, 90, 93-94

Q
Quadratic Weights, 20-22, 45, 89, 94, 104

R
Radical Weights, 88-89, 94, 103
Raw Scores, 30-32, 58, 78, 80, 96, 98, 100

S
SAS Enterprise Guide, x-xi, 4, 105-123
SAS Output, 36, 43, 54, 59, 62, 68, 72, 74-75, 91-93, 101-103, 109-110, 118
SAS/IML, x, 1, 88, 123
Scott Coefficient, 13, 25, 63-64
Scott's Generalized Coefficient, 87
SEED (see Initial Seed)
S-Index, 14, 16, 18
Standard Error, 2, 10, 33, 56,

59-61, 63-65
Unconditional, 90, 100
Statistical Significance, 66-67, 70, 93
Subject-Level Agreement, 80-81

T

TEST Statement, 66-67, 71
   KAPPA option, 67
   AGREE option, 71
Testing Kappa, 66-67, 71

U

U_P-value, 93
Unbalanced-Table Problem, x, 5-6, 37-38, 40, 42, 45, 56, 59, 121-122
   Solution, 38, 40, 42
Unconditional
   P-Value, 90, 93
   Standard Error, 90, 100

V

Variation of Raters' Classification Propensities 84
VCP, 84

W

WAP, 22-23
WCAP
   AC1, 27
   BP, 26
   Kappa, 22-24
   Scott's Pi, 25
Weighted AC_1, 104
   Agreement Probability 19
   Chance-Agreement Probability (see WCAP)
Weighting, 19
Weights
   Cicchetti-Allison (see CA-Weights)
   Fleiss-Cohen (See FC-Weights)
   Linear (See CA-Weights)
   Meaning, 50
   Quadratic (See Quadratic Weights)
   Radical (see Radical Weights)
Weigted Kappa, 22, 24
WKAPPA, 22
WT option, 46-47

Z

Z-Value, 60, 63-64

- For updates and more resources visit the AgreeStat page at:

  www.agreestat.com/agreestat.html.

- For PC SAS Users Download the macro programs **AgreeStat_2SAS.sas** and **AgreeStat_3SAS.sas**
- For Enterprise Guide users, download the following macro programs, and Enterprise Guide project files AgreeStat_2SASEG.sas, AgreeStat_3SASEG.sas AgreeStat_2EG.egp, and AgreeStat_3EG.egp